N. H. Seemayer · W. Hadnagy (Eds.)

Environmental Hygiene

With 69 Figures

Springer-Verlag
Berlin Heidelberg New York
London Paris Tokyo

Prof. Dr. NORBERT H. SEEMAYER
Medical Institute of Environmental
Hygiene at the University of Düsseldorf
Auf'm Hennekamp 50
4000 Düsseldorf 1, FRG

Dipl.-Biologe WOLFGANG HADNAGY
Medical Institute of Environmental
Hygiene at the University of Düsseldorf
Auf'm Hennekamp 50
4000 Düsseldorf 1, FRG

ISBN-13: 978-3-642-73768-8 e-ISBN-13: 978-3-642-73766-4
DOI:10.1007/978-3-642-73766-4

Library of Congress-Cataloging-in-Publication Data Environmental hygiene / N H. Seemayer, W. Hadnagy, eds p cm Includes index. 1 Environmental health–Congresses 2. Pollution–Congresses I. Seemayer, N H. (Norbert H), 1931- II Hadnagy, W (Wolfgang), 1941-
RA565.A2E577 1988 616 9'8–dc19 88-20134

Softcover reprint of the hardcover 1st edition 1988

Printing. Druckhaus Beltz, Hemsbach/Bergstraße
Binding: J Schaffer GmbH & Co KG., Grunstadt
2131/3130-543210 – Printed on acid-free paper

Preface

In a field as multifaceted as environmental research, the enormous amount of new data produced worldwide makes it difficult keeping up with the latest findings. Therefore, the "First European Meeting of Environmental Hygiene" was held in Düsseldorf on May 21 - 22, 1987 with the aim to bring together scientists from various sectors of environmental research and the intention of stimulating communication and contact between researchers. During the Meeting a wide range of topics were discussed, such as environmental pollution, exposure and health effects. The conference gave leading experts the opportunity to indicate important directions in the fast growing field of environmental research and to set priorities for further studies. In order to make the lectures of the Meeting available to scientists working in the field of environmental research, presented papers are published as a volume entitled "Environmental Hygiene".

We would like to express our special thanks to the speakers, the participants and all those who took part in organizing the Meeting. In particular, we are very grateful to the "Medizinisches Institut für Umwelthygiene" in Düsseldorf, and especially to the Director Prof Dr H.-W. Schlipköter for patronage of the Meeting.

Norbert H. Seemayer
Wolfgang Hadnagy

Contents

II. ANIMAL MODELS

III. BIOLOGICAL MONITORING

IV. EPIDEMIOLOGY

V. ENVIRONMENTAL CONTROL AND LEGISLATION

Contributors

You will find the addresses at the beginning of the respective contribution

INTRODUCTION

GENERAL ASPECTS OF ENVIRONMENTAL POLLUTION

Schlipköter, H.-W.

Medical Institute of Environmental Hygiene at the University of Düsseldorf, Auf'm Hennekamp 50,
D-4000 Düsseldorf, FRG

For more than 25 years, scientists all over the world are involved in studies on health effects of environmental pollution. These studies had become necessary by heavy air pollution in big cities and industrialized regions during the fifties. The level of an already high environmental pollution further increased during smog situations inhibiting transport and diffusion of air pollutants in the atmosphere. In those periods, concentrations of sulfur dioxide and particulates rose to values in the range of today's maximal threshold limits for working places (5 and 6 mg/m^3, respectively). During these smog episodes a dramatic increase of mortality was observed, especially in patients suffering from cardiovascular and lung diseases and in elderly people.

At the time, in the Ruhr District atmospheric concentration of carcinogenic polycyclic aromatic hydrocarbons was much higher than in most cities of Northern Europe and of the USA. To which extent these high concentrations of PAH are responsible for the increase of lung cancer mortality has not yet been resolved.

Caused by various efforts of environmental control, concentrations of atmospheric pollutants - except nitrogen oxides - became strongly reduced during the seventies. Simultaneously, however, the number of xenogenic substances in the environment further increased. Today there are more than 40,000 chemicals on the market, and several hundred new substances are added yearly by technical procedures. Many of them are released into the environment contaminating water, soil and air. They find their way to the human organism by food, drinking water, respiration, and through the skin. For example, in the city of Cologne more than 1,000 chemical substances have been detected in the atmosphere. So far, the health effects of all these substances, alone or in combination, are not well understood.

Therefore it is very important to investigate the effects of environmental pollutants on human health. For this purpose various methods and approaches can be used:

- In vitro studies on cells and cellular components
- in vitro studies on isolated organs
- animal experiments
- trials on volunteers and
- epidemiological studies.

In vitro experiments using mammalian cells and organ cultures may give first indications of a cytotoxic, mutagenic, and carcinogenic potential of a xenobiotic. They represent excellent screening systems having the advantage of short testing periods and low costs. However, they cannot entirely substitute experiments on an intact mammalian organism. For toxicological studies, more and more isolated organs are used permitting the elucidation of complex interactions between a chemical and its target organ. Furthermore, it is necessary to perform animal experiments, especially in cancer research and in order to investigate teratogenic and neurotoxic effects.

All these test systems may give valuable informations. However, concerning human health hazards caused by a xenobiotic, extrapolation of data to man remains a difficult task. An ultimate explanation can only be obtained by observations on exposed human populations. This can be achieved by experiments with voluntary persons, as well as by defined epidemiological studies of collectives with suspected or known exposition to environmental contaminants. However, experiments with voluntary persons need various restrictions, and epidemiological studies have to consider many factors which can influence results. It is also important that populations under investigation are exposed to numerous chemical substances simultaneously. Therefore, it is very difficult to correlate observed effects specifically to a single pollutant.

On the other hand, when using "biological monitoring" actual dimension of body burden with several pollutants can be estimated very well. Hereby, concentration of a particular pollutant can be determined in body fluids and tissues by seven excellent methods. For instance, atomic absorption spectroscopy is so sensitive, that the content of lead in the teeth shows as an indicator how long a child lived near to a lead plant or if his father is a lead worker or not. By this, sources of environmental pollution leading to body burden can be detected.

Each of the several methods mentioned can give valuable informations on effects of environmental pollutants on man. During this First European Meeting of Environmental Hygiene newest investigations in the field of environmental research will be presented and broad topics of interdisciplinary investigations in the field of environmental hygiene will be discussed. The aim of this meeting is, to bring together scientists from different fields for a productive exchange of ideas and newest results and informations.

I. IN VITRO TESTS

1. Cytotoxicity and Genotoxicity on Mammalian Cells

MECHANISMS OF O_3 AND NO_2 TOXICITY IN LUNG CELLS IN VITRO

Alink, G.M. and I.M.C.M. Rietjens
Department of Toxicology, Agricultural University, De Dreijen 12, 6703 BC Wageningen, The Netherlands

INTRODUCTION

Ozone and nitrogen dioxide are major toxic components of photochemical smog. The morphological, physiological and biochemical effects on the respiratory system of man and experimental animals have been investigated over the last decades.
The toxicity of ozone (O_3) and nitrogen dioxide (NO_2) is generally ascribed to their oxidative potential. Oxidation of cellular thiol groups and amino acids, and the oxidation of polyunsaturated fatty acids (PUFA) in the cell membrane offers a likely common mechanism of action. Many observations support a similarity in the toxic mode of action of O_3 and NO_2, as both gaseous compounds are known to have similar morphological, cytodynamical, functional and biochemical effects on the lungs of experimental animals. A similar mechanism for the reaction with unsaturated fatty acids is shown in figure 1.

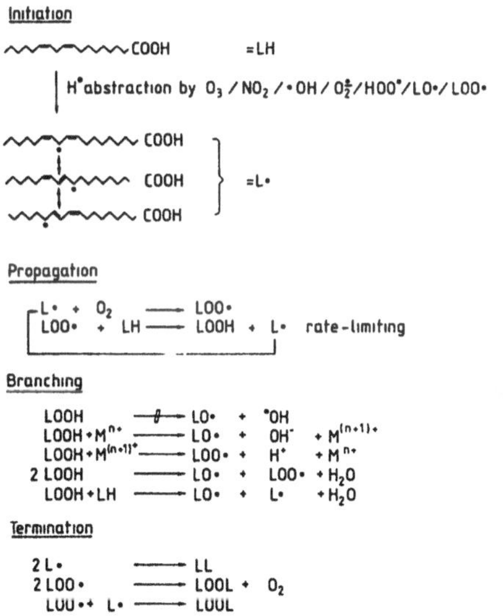

Fig. 1. Reaction scheme of radical mediated lipid peroxidation, initiated by hydrogen abstraction by ozone (O_3), nitrogen dioxide (NO_2) or other reactive oxygen species.

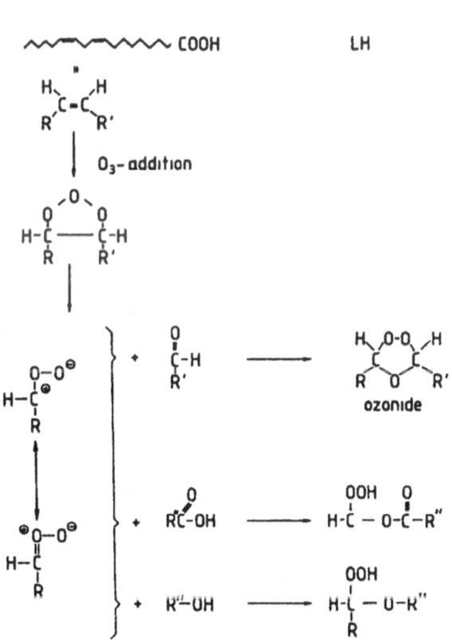

Fig. 2. The ionair Criegee mechanism for oxidation of unsaturated fatty acids initiated by ozone (O_3) addition.

O_3 and NO_2 may initiate hydrogen abstraction, which forms a lipid free radical, after reaction with oxygen lipid peroxide may be formed; a cascade of radical reactions may follow. Other results presented in literature provide evidence for differences in the mechanisms of action of O_3 and NO_2. They propose another mechanism for O_3 damage, namely that O_3 reacts with

PUFA by addition, following an ionair Criegee mechanism, giving rise to ozonides (Fig. 2).
However, the evidence for such dissimilarity, mainly arises from studies using biological fatty acid model systems, like liposomes and emulsions of fatty acids. It is clear that the mechanism of oxidative damage to unsaturated fatty acids in the cell membrane only can be studied adequately by using intact cell models. This consideration forms the basis for the present study.
To test possible differences or similarities in the mode of toxic action the following hypotheses were formulated. If a certain degree of cell damage caused by O_3 or NO_2 is initiated by the same reaction, the same reactive intermediates will be involved, while if different reactions occur, also different intermediates will be involved. Such a similarity (or dissimilarity) in reactive initial products may be reflected in the protection provided by cellular antioxidant systems.

MATERIALS AND METHODS

Cell Culture and Exposure Model

Cell cultures were exposed to O_3 and NO_2 by means of gas diffusion through a teflon film (1). Studies were performed with rat alveolar macrophages, isolated according to Mason et al. (3). Phagocytosis of yeast was used as a parameter for cellular damage.

Modification of Antioxidant Levels

The protection provided against equally toxic doses of O_3 and NO_2 by increased cellular levels of Vitamin C and Vitamin E was studied, as well as the effect of depletion of cellular glutathione (GSH) levels on the O_3 and NO_2 sensitivity of alveolar macrophages was compared. GSH was depleted with diethylmaleate and methionine sulfoxamine (4).

Measurement Membrane Fluidity

In order to demonstrate that the protection of Vit E was provided by radical scavenging and not by its structural effect on the cell membrane, the effect of protective amounts of Vit E on membrane fluidity was determined. In addition the effect of structural analogues of Vit E on membrane fluidity were tested (5).

Toxicity and Detoxification of Ozonides

Ozonides of methyl linoleate (MLO) have been synthesized to study their involvement in the mechanism of toxic action of ozone. The antioxidant protection and the spontaneous and enzyme catalyzed detoxification of methyl linoleate ozonide by GSH has been studied.

RESULTS AND DISCUSSION

In the in vitro exposure model used O_3 appeared to be about 10 times more toxic than NO_2. This resembles differences which have been found after exposure of experimental animals (2,5).
It was shown that preincubation of the alveolar macrophages with Vit E provided protection during a subsequent exposure to O_3 or NO_2. Vit E protected equally well against O_3 and NO_2 (Fig. 3) (5).
It was demonstrated that the Vit E protection is provided by its radical scavenging capacity and not by a changing of the membrane fluidity, as i) Vit E did not influence cell membrane fluidity at protective concentrations

and ii) structural analogues like the methylether of α-tocopherol (αT-OCH_3) and phytol could not offer any protection (5). Therefore it can be concluded that reactive intermediates scavenged by Vit E are important in the mechanism of both O_3 and NO_2 induced cell damage.

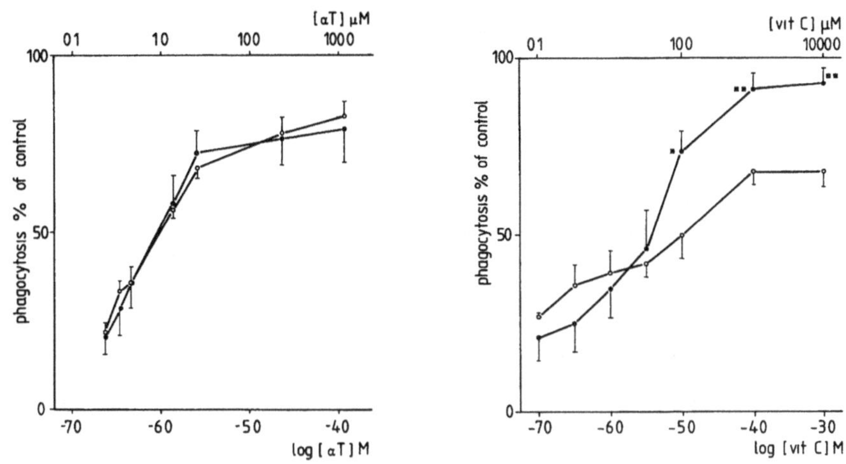

Fig. 3. Protection of alveolar macrophages against ozone (14-17 nmol O_3/disk) (-O-) and an equally toxic amount of nitrogen dioxide (172-177 nmol NO_2/disk) (-●-), provided by preincubation of the cells with Vitamin E (αT) or Vitamin C (Vit C).
* = $p < 0.05$, ** = $p < 0.01$ (unpaired Student's t-test)

However, other results clearly indicated that the kind of radicals and/or reactive intermediates, and thus the toxic reaction mechanism involved, must be different in O_3 and NO_2 induced cell damage. This was concluded from the following observations. First, preincubation of the cells with Vit C provided better protection against NO_2 than against O_3 (Fig. 3).
Secondly, GSH-depletion affected the cellular sensitivity towards O_3 to a greater extent than towards NO_2 (5).
One of the possibilities to explain these results might be the involvement of lipid (peroxyl) free radicals, formed in a radical mediated peroxidative pathway in NO_2 induced cell damage, while O_3 induced cell damage may proceed by lipid ozonides.
More evidence on the involvement of ozonides in the mechanism of toxic action of ozone have recently been obtained in studies with ozonides of methyl linoleate (6). The ozonide of methyl linoleate is 1000 x more toxic than methyl linoleate. The pattern of antioxidant protection is the same as for ozone, e.g. Vit E protects AM against MLO, glutathione-depletion enhances the sensitivity of cells for MLO.
Figure 4 shows the detoxification of methyl linoleate ozonide by GSH. Furthermore it is shown that glutathione S-transferase (GSHTr) can catalize the GSH mediated detoxification of ozonide. Preincubation of ozonides with both GSH and GSHTr completely abolished its toxicity. That the detoxification of ozonides by GSH is catalysed by GSHtr offers a new viewpoint on the role of GSH in the protection of the cell against ozone. The similarity in antioxidant characteristics between ozone and ozonides, and the differences observed in mechanisms of toxic action between O_3 and NO_2,

indicate that the addition or Criegee mechanism as shown in artificial model systems is also a possible mechanism for the toxicity of O_3 in the intact cell.

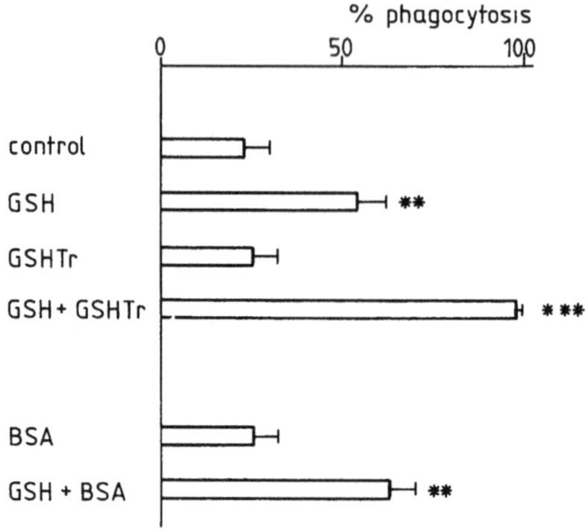

Fig. 4. Detoxification of methyl linoleate ozonide (86 μM) by preincubation of the ozonide with GSH (1 mM), GSHTr (0.01 mg/ml) or both. Control incubations with bovine serum albumin (BSA) were carried out using 0.10 mg BSA/ml.

REFERENCES

(1) Alink, G.M., J.C.M. van der Hoeven, F.M.H. Debets, W.S.M. van de Ven, J.S.M. Boley and J.H. Koeman (1979) A new exposure model for in vitro testing of effects of gaseous pollutants on mammalian cells by means of gas diffusion through plastic films, Chemosphere, 2, 63-73.

(2) Evans, M.J., L.V. Johnson, R.J. Stephans and G. Freeman (1976) Renewal of the terminal bronchiolar epithelium in the rat following exposure to NO_2 or O_3, Lab. Invest., 35, 246-257.

(3) Mason, R.J., M.C. Williams, R.C. Greenleaf and J.A. Clements (1977) Isolation and properties of type II alveolar cells from rat lung, Am. Rev. Resp. Dis., 115, 1015-1026.

(4) Rietjens, I.M.C.M., G.M. Alink and R.M.E. Vos (1985) The role of glutathione and changes in thiol homeostasis in cultured lung cells exposed to ozone, Toxicology, 35, 207-217.

(5) Rietjens, I.M.C.M., M.C.M. Poelen, R.A. Hempenius, M.J.J. Gijbels and G.M. Alink (1986) Toxicity of ozone and nitrogen dioxide to alveolar macrophages: comparative study revealing differences in their mechanism of toxic action, J. Toxicol. Environ. Health, 19, 555-568.

(6) Rietjens, I.M.C.M., H.H. Lemmink, G.M. Alink and P.J. van Bladeren (1987) The role of glutathione and glutathione S-transferases in fatty acid ozonide detoxification, Chem.-Biol. Interactions, 62, 3-14.

A STUDY ON THE BIOLOGICAL EFFECTS OF POWER STATION ASH

R.C.Brown, *A.Poole and **C.J.Turver
MRC Toxicology Unit, Woodmansterne Road,Carshalton Surrey.
* SKF Research Ltd, The Frythe, Welwyn, Herts.**Dept. of Microbiology, The University, Leicester

INTRODUCTION.

The possible adverse effects of emissions from coal burning power stations been the subject of several studies (1).In British power stations coal is mainly burnt as a freshly milled powder (pulverized fuel) and most of the resulting ash is collected by electrostatic precipitators. The collected material has a number of commercial uses and is a valuable by-product of power generation. A small quantity of the ash is not precipitated and normally escapes to the atmosphere, this material consists mostly of the finest particles and is sometimes (as here) refered to as stack ash.

The biological activities of the stack ash and samples of the ash from hoppers containing the electrostatically precipitated ash have both been examined.

MATERIALS AND METHODS

Ash samples: These were provided by the Central electricity Generating Board of the United Kingdom (CEGB). The stack ash samples were collected from the effluent gases at three power stations . Four samples of precipitated ash were taken from the same three power stations. The fourth sample being collected at one power station, and tested within 24h of collection. All samples were stored in the dark at -60C under nitrogen. The ash was sterilized before use by autoclaving at 121C for 15 minutes. They were then suspended in tissue culture medium by gentle sonication.
Cells and Cytotoxicity assays Three cell lines were used in this work. Growth conditions and experimental details of the assays have been published previously (2,3,4).

RESULTS

Physical properties of the dust: All the samples were brown/grey powders readily dispersable in aqueous solutions. The stack ash samples were examined by electron microscopy and the number of particles in unit mass was estimated this information is given in Table 1. The mean diameters of all the stack ash samples were similar (approx 1.5um) though there were more of the larger particles in the A with fewest in the C.
Interaction with cells As it was was uncertain as to whether these materials could be used in tissue culture some preliminary experiments were conducted in which suspensions were incubated with cells and examined by electron microscopy. Ash particles were seen in and on the cell in many types of association.

Table 1. Particle number in the Stack Ash Samples

Power station	Total particle numbers in 1 ng	
A	20.8	21.4
B	45.7	76.3
C	194.1	108.0

The numbers are given for both samples from each power station.

Cytotoxicity The stack ash samples were examined for their cytotoxicity. In all cases samples from power station C were the most toxic with LD50s of less than 1/2 that for the other dusts. Examination of the precipitated ash samples showed that the fresh sample and that from power station B were much the most active (Table 2).

Table 2.
Toxicity of precipitated ash samples to C3H10T1/2 cells

Dose ug/ml	A	B	C	Fresh
15	96	56	94	69
30	80	56	59	53
60	59	13	65	19

Cell survival is expressed in the table as a percentage of that in untreated control cultures.

Arachidonic acid release In tables 3 and 4 it can be seen that the stack ash samples had very low activity in this system and these results are discussed below. The precipitated ash samples were if anything even less active than the stack ash, only the "fresh sample" produced any significant release and only at the highest concentration (table 4).

Table 3.
Arachidonic acid release from P388D1 cells Stack ash.

Dose	A 1	A 2	B 1	B 2	C 1
64	4.09	3.80	7.41*	4.52	4.55
	(0.37)	(0.05)	(1.03)	(0.37)	(0.84)
125	4.50	4.96	5.32	5.64	4.72
	(0.60)	(0.48)	(0.29)	(0.73)	(0.55)
250	4.61	5.84	6.11	6.69	4.85
	(0.50)	(0.52)	(0.46)	(0.75)	(0.43)
500	7.2*	8.11*	6.12	5.63	6.46
	(1.52)	(1.26)	(1.46)	(0.3)	(0.37)
Control	4.85	(1.34)			

Only one of the samples from station C was tested the figures in brackets are the 95% confidence limits.

Table 4. Arachidonic acid release Precipitated Ash Samples

Dose	A	B	C	"Fresh"
64	4.25	4.54	4.49	4.29
	(0.10)	(0.25)	(0.51)	(0.41)
125	4.25	4.84	4.62	4.47
	(0.09)	(0.32)	(0.61)	(0.14)
250	4.22	4.33	4.34	4.47
	(0.12)	(0.33)	(0.04)	(0.18)
500	4.51	4.91	4.31	5.44*
	(0.24)	(0.51)	(0.25)	(0.33)
Control	4.21	(0.30)		

DISCUSSION

There may be some correlation between the pathogenicity of a dust *in vivo* and its cytotoxicity *in vitro* and this has been suggested for various ash-like materials (5,6,7). Srivistava & Mishra (8,9) studied the lipid in lung and other organs and suggested that the changes they observed could be related to the *in vivo* cytotoxicity of coal fly ash.

This observation and the suggestions by Lantz & Hinton (10,11) that coal fly ash was pathogenic suggest a possible causal link between the mechanism of pathogenesis *in vivo* and cytotoxicity *in vitro*.

Garrett *et al* (5), using a reduction in ATP as an indication of cytotoxicity found that a "model fly ash" was relatively non-toxic to rabbit alveolar macrophages at 1000 ug/ml. This was in agreement with the findings of Hill & Hobbs (12) who also reported that coal fly ash was not cytotoxic.

A better approach to the use of cytotoxicity is to use an endpoint which may have a greater causal relationship to pathogenesis. One such endpoint is increased prostaglandin which is a common response to cell damage or phagocytosis and has been reported as one biological activity of mineral dusts. For example we have demonstrated that asbestos can cause the activation of phospholipase A2 and the release of prostaglandin E2 from macrophage like cells (3).

Two of the samples of stack ash and the fresh precipitated ash sample produced significant release of label at the highest concentration used; one other value was significantly greater than the control. Even titanium dioxide, generally regarded as biologically inert, showed more activity than the ash samples at the lower concentrations. In brief the ash samples are the most inert materials that we have so far examined for this activity.

In vivo studies undertaken with ash produced in other countries have found that both precipitated and stack ash were fibrogenic. Several such studies also support the view that ash from fluidised bed burners will prove to be more active than that from conventional sources. The low activity we have found in power station ash suggests that it is not likely to pose a serious health hazard. We will publish elsewhere the results of genotoxicity studies on these materials, though here again the results are mainly negative.

SUMMARY AND CONCLUSIONS

Ash collected at British power stations was examined for its activity in a variety of tests for cytotoxicity. While this material was slightly cytotoxic it is one of the most inert substances yet examined in our laboratory. It is unlikely that this type of material would pose any health hazard. Such conclusions may not apply to ash produced under different combustion conditions.

REFERENCES

1. Fisher G.L. (1983) Biomedically relevant chemical and physical properties of coal combustion products. Environ.Health Perspec. 47, 189-199
2. Chamberlain, M. & Brown, R.C. (1978) The cytotoxic effects of asbestos and other mineral dust in tissue culture cell lines. Br.J.Exp.Path, 59, 183-189.
3. Brown, R.C. & Poole, A. (1984) Arachidonic acid release and prostaglandin synthesis in a macrophage like cell line exposed to asbestos. Agents and Actions. 15, 336-340
4. Brown R.C. Poole A. & Flemming, G.T.A. The influence of asbestos dust on the onchogenic transformation of C3H10T1/2 cells. Cancer Letters 18, 221-227
5. Garrett N.E. Campbell J.A. Stack H.F. Waters M.D. & Lewtas J. (1981a) The utilisation of the rabbit alveolar macrophage and CHO cell for the evaluation of the toxicity of particulate materials Environ. Res. 24, 345-365
6. Garrett N.E. Campbell J.A. Stack H.F. Waters M.D. & Lewtas J.(1981b) The utilisation of the rabbit alveolar macrophage and CHO cell for the evaluation of the toxicity of particulate materials 2.Particles from coal related processes. Environ Res.24, 366-376
7. Garrett N.E. Campbell J.A. Stack H.F. Jackson M.A. & Lewtas J. (1983) Cellular toxicity in chinese hamster ovary cell cultures. Environ.Res. 32, 466-473
8. Srivistava P.K. & Misra U.K. (1985) Cytotoxicity of intratracheally administered coal fly ash: studies on lipids in the lung of rats. Arch.Environ.Contam.Toxicol. 14, 95-103
9. Srivistava P.K. & Mishra U.K. (1985) Cytotoxicity of intratracheally administered coal fly ash: studies on liver and plasma lipids in rats. Arch.Environ Contam Toxicol 14, 105-110
10. Lantz R.C. & Hinton D.E. (1984) Pulmonary toxicity associated with fly ash from fluidized bed coal combustion. Toxicol Appl Pharmacol 75, 44-51
11. Lantz R.C. & Hinton D.E. (1986) Developments of alterations in hamster distal lung following exposure to fly ash from fluidized bed coal combustion: A morphometric study. Toxicol appl Pharmacol 82, 132-139

INDUCTION OF MALIGNANT TRANSFORMATION IN HAMSTER FETAL CELLS (HFC) BY DIVERSE ENVIRONMENTAL CARCINOGENS

DiPaolo, J.A., J. Doniger and N.C. Popescu
National Cancer Institute, NIH, Bethesda, Maryland, 20892, U.S.A.

INTRODUCTION

Toxicity of environmental agents, more specifically cancer risk, is important for estimating the human health risk associated with exposure to substances present in foods, drugs, consumer products and industrial environments. Epidemiologic studies as summarized by the IACR incriminate 23 chemicals as human carcinogens (1). These studies correlate with long term animal carcinogenesis studies. For several chemicals, aflatoxin B_1, 4-aminobiphenyl, bis(chloromethyl) ether, diethylstilbesterol, mustard gas, and vinyl chloride, animal studies predated epidemiologic evidence. Concentration and duration of exposure are important for estimating the effect of chemicals on humans. Results of animal studies indicate that protracted exposure to a weak carcinogen may be just as deleterious as a pulse exposure to a potent carcinogen, factors that must be evaluated in making regulatory decision based on predictive tests.

The advantage of utilizing mammalian cells _in vitro_ for carcinogen assay and for mechanism studies are many. First of all, _in vitro_ data agree with animal studies and correlate positively with the epidemiologic studies. The HFC model permits quantitation: a linear dose response (2). HFC share with human cells a stable chromosome constitution and senescence. Spontaneous transformation is a rare phenomenon. Just as important, with the HFC transformation is induced, not due to selection of preexisting neoplastic cells, and it is direct because it does not require viral activation (3).

RESULTS AND DISCUSSION

Epidemiological evidence has indicated a relationship between exposure to sunlight and incidence of skin cancer in humans. Humans with xeroderma pigmentosum are highly sensitive to UV-induced skin cancer. A series of action spectra for transformation, lethality, and pyrimidine dimer (PD) formation in HFC when normalized to 254 nm were superimposable on the average action spectrum for microbial organisms in terms of mutagenesis, survival and PD formation (4). Because DNA is the target for PD production and because of the similarities between the transformation and cellular PD action spectra, DNA must also be the target for UV-induced carcinogenesis (5). Survival of HFC after UV was similar to that of human skin fibroblasts which excise PD efficiently. Postreplication repair cannot account for the ability of HFC to survive because it is slower than in human fibroblasts. Therefore, other repair mechanisms may be responsible for HFC survival and perhaps transformation. However, experiments utilizing human cells have not resulted in neoplastic transformation (6). Alkalating agents which mimic certain properties of X-irradiation are also capable of transforming HFC (7). For example, N-methyl-N'-nitro-N-nitrosoguanidine (MNNG), N-methyl-N-nitrosourea (MMS) or methyl methanesulfonate induced dose-dependent transformation with MNNG are most effective. Comparison of induction repair of O^6 and N^7methylguanidine at concentrations that induce equivalent transformation frequencies demonstrated that the induction of O^6-methylguanine was the same for all carcinogens whereas N-methylguanine induction was highest with the least potent carcinogen, MMS. Neither the rate of DNA replication nor the size distribution of nascent strands correlated with O^6-methyl-guanine induction. These results support the hypothesis that O^6-methylguanine is a critical lesion for initiation of carcinogenesis by methylating agents. In rats treated with ethylnitrosurea, N^7 was removed efficiently from both the brain and liver, but O^6-methylguanine was efficiently removed from the liver but not from the brain where tumors developed. The O^6-alkylguanine is considered promutagenic because it induces base transitions during DNA replication, whereas N^7-alkylguanine does not. These results imply that initiation is due to O^6-alkylguanine lesions. The <u>in vitro</u> results provide additional evidence implicating O^6 as a critical lesion for initiation by methylating agents. The increased transformation frequency due to O^6-methylguanine relative to the mutation frequency suggests that the target size for initiation of carcinogenesis is much larger than for mutagenesis. Furthermore, analysis indicates that one of many genes can initiate carcinogenesis or that initiation is not simply the result of a single base mutation.

Two chemical environmental agents relevant to problems that may affect humans are bisulfite and asbestos. Bisulfite is a food and pharmaceutical preservative, is ubiquitous in the form of SO_2 as a pollutant and is suspected to involve a known molecular phenomenon often associated with mutagenesis. It is a decay product of organic material and results from combustion of coal and petroleum. The epidemiologic evidence suggests an association between atmospheric SO_2 with cardiac and respiratory illnesses and lung cancer. Whereas at acid pH SO_2 causes deamination of cytosine residues that are potentially mutagenic in <u>E. coli.</u>, at neutral pH and at nontoxic concentrations it neither induces tryptophane revertants in <u>E. coli</u> or causes mutation to ouabain-resistance in V79 cells (8). In mammalian cells as well as in <u>E. coli</u>, bisulfite plus ultraviolet radiation causes enhanced mutagenesis. In HFC, bisulfite causes morphologic transformation which leads to neoplastic transformation; however, bisulfite did not increase UV associated transformation (9). In terms of acute effects bisulfite produces minimal changes in DNA. Neither postreplication repair or induction of replication incision repair was noted. Moreover, concentrations that produced transformation had a minimal effect on induction of sister chromatid exchanges. Malignant lines were aneuploid with structural alterations similar to those identified in cells transformed by carcinogens that produce extensive DNA damage. Analysis of DNA synthesis after treatment with 0-50 mM bisulfite indicates an inhibition of semi-conservative DNA synthesis probably as a result of a reduction in the number of functioning replicons (10). Examination of a number of malignant cell lines by 2D electrophoresis showed that both qualitative and quantitative polypeptide changes occurred. All malignant lines, regardless of chromosome constitution, exhibited four qualitative polypeptide changes. One of these, a new polypeptide, correlated with tumorigenicity (11).

Epidemiologic evidence has demonstrated an association between exposure to asbestos and mesothelioma of the pleura and peritoneum. Lung cancer incidence among asbestos workers who also are cigarette smokers is 92 times that of the general population. A further comparison to nonsmokers or those who do not work with asbestos is indicative of a cocarcinogenic effect. Different asbestos fibers can induce morphologic transformation in HFC which results in tumorigenicity. The exposure of cells to benzo(a)pyrene (BP) or UV and different concentrations of asbestos results in transformation enhancement only by BP (12). The enhancement with BP was dose-dependent with all fiber species used except for amosite which was dose-independent. Attempts to reproduce this phenonomen *in vitro* using human cells have not been successful to date. The cytogenetic examination of human malignant mesotheliomas that *in vitro* appear either epithelial-like or fibroblast-like exhibited nonrandom alterations involving chromosomes 1, 3 and 7 (13). The breakpoint of 3 is usually at 3p21-14 and is associated with interstitial or terminal translocations and deletions. A number of other types of cancer also involve the same breakage point. Whether breakage at this specific site is involved in induction and/or progression leading to mesothelioma is an open question. There is also a possibility that this region on 3p may possess a suppressor gene the removal of which may trigger the process leading to cancer.

To understand the underlying mechanism of the cancer process, it is necesssary to identify the cancer genes. The search for such genes and for mechanisms that generate them must take into consideration that, at the cellular level, cancer is a very rare event. The HFC model is particularly suitable for studying how specific genes are involved in carcinogenesis by environmental chemicals because the induction of transformation occurs without viral activation. With oncogenes it is possible to determine both the changes responsible for their activation and to determine when they occur or are expressed. It is obvious that most carcinogens cause DNA damage of hamster cells. The transfection of DNA derived from a number of transformed lines induced foci on 3T3 cells and furthermore gave second round transformation while normal HFC DNA was negative. Analysis for known *ras* or *abl* genes have been negative. When a large number of transformed lines, approximately 40, were examined by radioimmune assay for *ras* genes, no mutant p21 was found. Thus, it appears that if oncogenes are involved in the transformation of hamster cells to the malignant state, hamster oncogenes must represent a new class of oncogenes that have not been heretofore described.

SUMMARY AND CONCLUSIONS

A rationale exists for the study of cancer induction at the cellular level. Transformation of HFC can be quantitated; frequency follows a linear relationship with dose phenomenon and is due to one hit kinetics. Transformed colonies produce cell lines which produce tumors. In vitro asseys utilizing mammalian cells have made possible the rapid evaluation of carcinogenicity of environmental agents. The addition of a host-mediated step decreases the possibility of false negatives, which may result from the requirement for metabolic activation of the chemical. Thus in vitro studies described have a high probability of providing practical methods for determining which chemicals have a potential of producing cancer. Furthermore, the nature of the cell-target insult interaction, the nature of the ultimate carcinogen, the degree to which any agent acts alone, be it a chemical, a virus, or irradiation, and the extent to which one agent interacts with another from the same or different category can be determined.

REFERENCES

(1) Wilbourn, J.D., L. Haroun, H. Vaino, et al (1984) Identification of chemicals carcinogenic to man. Toxicol. Pathol., 12, 397-399.

(2) Gart, J.J., J.A. DiPaolo and P.J. Donovan (1979) Cancer Res., 39, 5069-5075.

(3) Reitz, M.S., W.C. Saxinger, R.C. Ting, R.C. Gallo and J.A. DiPaolo (1977) Lack of expression of type C hamster virus after neoplastic transformation of hamster embryo fibroblasts by benzo(a)pyrene. Cancer Res., 37, 3585-3589.

(4) Setlow, R.B. (1966) Cyclobutane-type pyrimidine dimers in polynucleotides. Science, 153, 379-386.

(5) Doniger, J., E.D. Jacobson, K. Krell and J.A. DiPaolo (1981) Ultraviolet light action spectra for neoplastic transformation and lethality of Syrian hamster embryo cells correlate with spectrum for pyrimidine dimer formation in cellular DNA. Proc. Natl. Acad. Sci., 78, 2378-2382.

(6) Greiner, J.W., C.H. Evans and J.A. DiPaolo (1981) Carcinogen-induced anchorage-independent growth and in vivo lethality of human MRC-5 cells. Carcinogenesis, 2, 359-362.

(7) Doniger, J., R.S. Day and J.A. DiPaolo (1985) Quantitative assessment of the role of O^6-methylguanine in the initiation of carcinogenesis by methylating agents. Cell Biology, 82, 421-425.

(8) Mallon, R.G., and T.G. Rossman (1981) Bisulfite (SO_2) is a comutagen in E. coli and C. hamster cells. Mutat. Res., 88, 125-133.

(9) DiPaolo, J.A., A.J. DeMarinis and J. Doniger (1981) Transformation of Syrian hamster embryo cells by sodium bisulfite. Cancer Letters, 12, 203-208.

(10) J. Doniger, R. O'Neill and J.A. DiPaolo (1982) Neoplastic transformation of Syrian hamster embryo cells by bisulfite is accompanied with a decrease in the number of functioning replicons. Carcinogenesis, 3, 27-32.

(11) Wirth, P., J. Doniger, S.S. Thorgiersson and J.A. DiPaolo (1986) Altered polypeptide expression associated with neoplastic transformation of Syrian hamster cells by bisulfite. Cancer Res., 46, 390-399.

(12) DiPaolo, J.A., A.J. DeMarinis and J. Doniger (1983) Asbestos and benzo(a)pyrene synergism in the transformation of Syrian hamster embryo cells. Pharmacology, 27, 65-73.

(13) Popescu, N.C., S.C. Amsbaugh, P.A. Chahinian and J.A. DiPaolo (1987). Nonrandom deletions of chromosome 3 in human malignant mesothelioma. Proc. of the Seventy-Eighth Annual Meeting of the AACR, 28, 37.

BIOCHEMICAL AND MORPHOLOGICAL STUDIES ON OZONE AND NITROGEN DIOXIDE TREATED CULTURED LUNG CELLS

Ebert, W., E. Moll, D. Komitowski* and I. Vogt-Moykopf
Clinic for Thoracic Diseases, Amalienstr. 5,
D-6900 Heidelberg, FRG
*German Cancer Research Center, D-6900 Heidelberg, FRG

INTRODUCTION

Acute or chronic exposure to air pollutants such as O_3, NO_2, and SO_2 results in a sequence of damage, inflammatory response, proliferation, and repair in the lung.
A variety of cells of the bronchial tree and the alveoli such as mucus secreting cells, ciliated epithelial cells, clara cells, type I and type II cells as well as endothelial cells were shown to be injured (1). Most studies, however, were done on animal models by histological examinations and biochemical measurements of whole tissue sections. In order to study the effect of air pollutants on definite cell populations of the human lung we developed an in vitro model allowing the morphological and biochemical investigation of directly gassed cells.

MATERIAL AND METHODS

Primarily cultivated human lung cells and cell lines from the pulmonary artery endothelium (bovine, CPAE) as well as cell lines derived from human alveolar type II cells (A 549, both lines from American type culture collection) are grown in cell culture media containing 20% fetal calf serum. They are injected into heatable cell cultivation chambers (37°C, modified Pentz-chambers, Heraeus, Hanau - initial cell density: 1.5 X 10^5 cells / cm^2 of growth surface). Control specimen were gassed with synthetic air containing 5% CO_2. O_3 was produced by an ozone generator (Tecan, München), certified gas mixtures of 80ppm NO_2 in synthetic air / 5% CO_2 were purchased from Messer-Griesheim (Frankfurt). Lactate dehydrogenase (LDH) activity was determined using a kit of Boehringer-Mannheim. The trypan blue exclusion test served as an additional assay for measuring cell injury. Cell preparations for scanning and transmission electron microscopy were performed as described (2). In order to determine morphological alterations due to O_3-exposure treated and untreated CPAE-cells were fixed according to Carnoy (3) for 20 min. and Feulgen-stained (DNA-staining, 3). 400 Feulgen-stained cell nuclei of a test or control series were analyzed by means of a digital image processing system based on light microscopic screening with complementary light (550nm, 3). Statistical evaluations were performed by t-tests.

RESULTS

Quantification of cell damage

Exposure of endothelial cells to O_3 for 72 hours results in a distinct increase of trypan blue-stained cells and of LDH-release in relation to the control. This increase correlates with ozone concentration. 0.1ppm O_3 are able to cause cell damage (fig. 1). Using a constant O_3-concentration of 0.3ppm the extent of cellular damage depends on the time of gas exposure (fig. 2). Type II-cells are more resistant to O_3-exposure than endothelial cells. If treated with 0.3ppm O_3 for 72 hours, the increase of type II-cell damage relative to control assays only amounts to 12% (trypan blue-stained cells), respectively 49% (LDH-activity) of the data obtained by using endothelial cells. As demonstrated in figures 3 and 4 treatment of in vitro-cultivated lung cells with NO_2 also causes cellular injury in a time-and dose-dependent manner. The minimum NO_2-concentration to cause measurable cell damage of endothelial or type II-cells is between 1.25 and 5ppm NO_2 (figs. 3,4).

Electron microscopic observations

In SEM-preparations primarily cultured human lung cells and CPAE cells gassed for 20 to 72 hours with 95% oxygen or 0.3ppm ozone have an elongated filamentous or rounded and clumped appearance whereas control cells are flattly spread on the glass surface. Transmission electron microscopic investigations of O_3-treated endothelial cells in comparison to control cells demonstrate ultrastructural alterations. Some of the cell nuclei are found to be irregularly shaped and numerous large autophagosomes are prominent. The majority of cells, however, show more subtle modifications: The endoplasmic reticulum and mitochondria look swollen, the nucleolar matrix appears to be more condensed and the chromatin structure shows a more coarse distribution.

Quantification of chromatin rearrangements

By means of light microscopic digital image analysis some characteristic alterations of intranuclear structure are quantified. Endothelial cells exposed to 0.3ppm ozone for 20 hours show enlarged nuclei in comparison to control cells. The mean square dimension of nuclear cross sections has augmented by 24%, the mean nuclear circumference by 12.5%. The optical density of nuclei corresponding to DNA-density, however, is decreased (fig. 5). The number of local density maxima corresponding to heterochromatin areas increased in the nuclei of gassed cells being partially redistributed to the nuclear periphery (fig. 5).

SUMMARY AND CONCLUSIONS

Gassing of isolated lung cells in modified Pentz-chambers is an in vitro model suitable for the examination of cellular damage by air pollutants. The extent of cell injury correlates with the concentration of O_3 or NO_2 and time of exposure. Minimum effective concentrations for impairment of lung endothelial cells were found to be 0.05 to 0.1ppm O_3 or 1.25 to 5ppm NO_2. These data are in the range of maximum burden of immissions measured in town centers (4). Electron microscopic examinations reveal elongation or rounding of gassed cells, which corresponds to a partial retraction from the glass substrate. These effects may reflect membrane injury, presumably due to lipid peroxidation. Nuclear deformation and abundance of autophagosomes indicate severe cell damage, i.e. autolysis. Less dramatic ultrastructural alterations like swelling of mitochrondria and endoplasmic reticulum may arise from general impairment of cell metabolism, whereas condensation phenomena of nucleoli and chromatin are hints to possible disturbance of ribosomal synthesis and gene activity. On a light microscopic level cells exposed to ozone show redistribution of euchromatin and heterochromatin which also supports an altered gene expression. Special biosynthetic pathways (e.g. those for xenobiotic metabolism) may be activated by irritant gases. The observed enlargement of nuclear size apparently does not include an increase of total DNA-content. More detailed in vitro studies in correlation with in situ-animal experiments will further have to clarify the effects of air pollutants on human lungs.

REFERENCES

(1) Gail, D.B. and C.J.M. Lenfant (1983) Cells of the Lung: Biology and Clinical Implications, Am. Rev. Resp. Dis. 127, 366-387.

(2) Werling, H.-O., E. Spiess, P. Aulenbacher and N. Paweletz (1985) Adhesion and Spreading Characterization of a Rat Tumor Cell System Exhibiting Different Metastastic Behavior, Invasion Metastasis 5, 270-294.

(3) Komitowski, D. and G. Zinser (1985) Quantitative Description of Chromatin Structure During Neoplasia by the Method of Image Processing, Analyt. Quant. Cyt. Hist. 7, 178-182.

(4) TA Luft (1983) and Allgemeine Forstzeitschrift 7 (1984).

Figs. 1-4:
Measurements of cell damage of pulmonary artery endothelial cells (CPAE) or alveolar type II-cells (A 549) after exposure to O_3 or NO_2. The increase of the portion of trypan blue permissive cells and of lactate dehydrogenase (LDH) activity is plotted relative to the mean values of controls (air / 5% CO_2).

Fig. 5:
Alteration of CPAE cell nuclear parameters after exposure to 0.3ppm ozone for 20 hours. Data is received by digital light microscopic image analysis. Mean values of test and control data (C) differ significantly ($p < 0.001$, t-test).

Fig 1 Exposure of pulmonary endothelial cells to ozone for 72 hrs

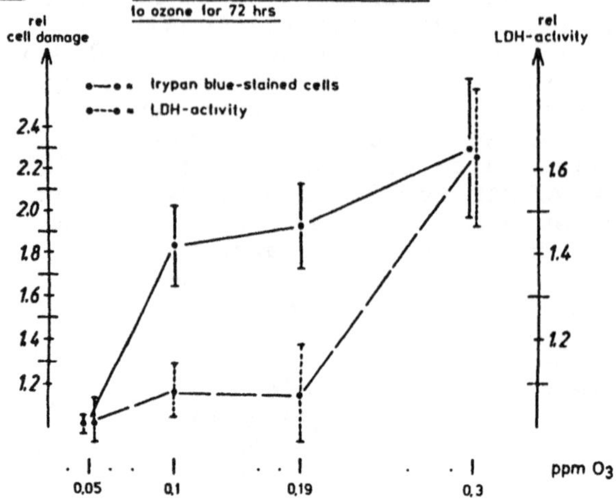

Fig 3

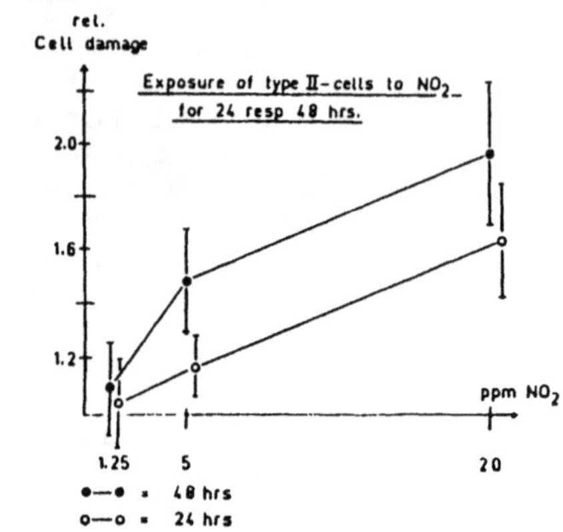

Fig 2 Exposure of pulmonary endothelial cells to 0,3ppm ozone

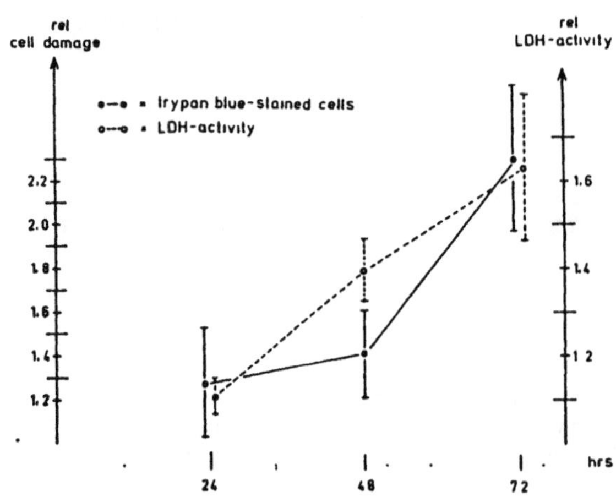

Fig 4

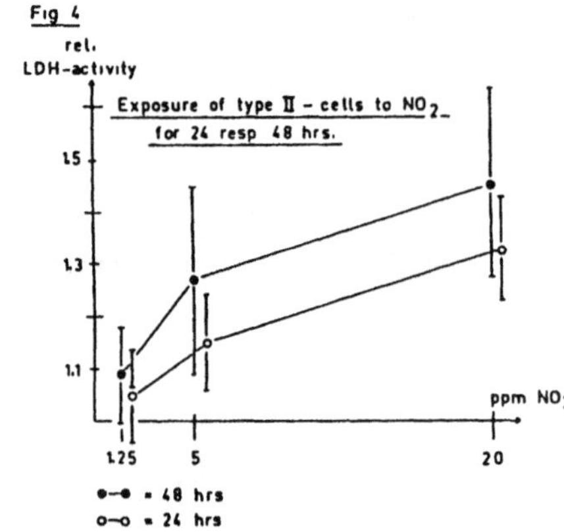

Fig 5

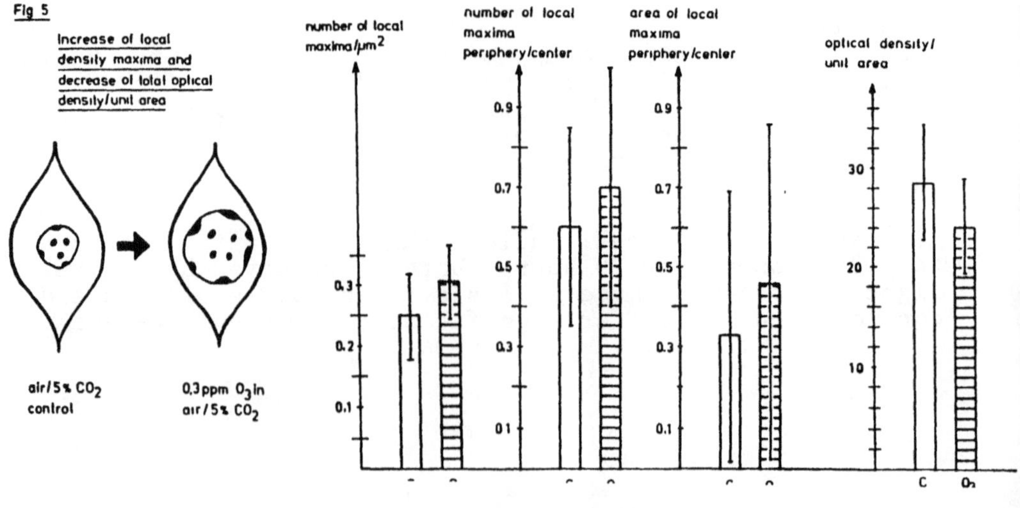

INDUCTION OF SISTER CHROMATID EXCHANGES BY ASBESTOS FIBRES IN COMBINATION WITH OTHER MUTAGENS STUDIED IN CHINESE HAMSTER CELLS IN VITRO

Fischer, A.B.* and J.L. Kaw°
Institute of Hygiene, University of Giessen, Friedrichstr. 16, D-6300 Giessen, FRG*; Industrial Toxicology Research Centre, Post Box No. 80, Lucknow-226001, India°

INTRODUCTION

The naturally occurring mineral asbestos is known to constitute an occupational as well as environmental health hazard leading to lung fibrosis and two types of cancer: mesothelioma and pulmonary carcinoma. Smoking potentiates the risk of lung cancer. The responsible components of cigarette smoke have so far not been identified, though polycyclic aromatic hydrocarbons have been suspected.

In mammalian cell cultures *in vitro* asbestos fibres but also Man-Made Mineral Fibres are cytotoxic. In addition, morphologic cell transformation has been demonstrated for chrysotyle, crocidolite, amosite, and erionite (1, 2). These fibres also produce cytogenetic damage such as above all changes of chromosome numbers and ploidy as well as structural alterations (gaps, breaks and exchanges) (3, 4, 5). However, studies on the induction of point mutations have usually been negative with the exception of the investigation by Huang et al. which reported weakly positive effects of amosite, chrysotile and crocidolite (6,7). Studies using sister chromatid exchanges (SCE) as a mutagenic endpoint have to date yielded conflicting results. While three working groups found no effects in human lymphocytes and lymphoblastoid cells, C3H mouse cells, and Chinese hamster cells V79, chrysotile, crocidolite, and erionite were positive in Chinese hamster ovary cells (8 - 12)

The present study had the purpose to reexamine the induction of SCE by asbestos fibres and to see if a synergistic effect of other environmental mutagens could be detected. The following mutagenic xenobiotics were employed: methanesulfonic acid, ethyl ester (EMS) as a model mutagen, benzo(a)pyrene (BaP) as an important aromatic constituent of cigarette smoke, and cadmium, because this heavy metal also occurs in cigarettes and has recently been shown to induce lung cancer in rats (13).

MATERIALS AND METHODS

Cells. Chinese hamster peritoneal cells, line B14F28, with a mean chromosome number of 22 - 23 were cultured in MEM supplemented with non-essential amino acids and 5 % Newborn Calf Serum. *Fibres.* Modified UICC asbestos was used: chrysotile (CFF; 80.6% 2µm, 99% 8 µm) and crocidolite (KFF; 98.9% 5 µm). *Mutagens.* Benzo(a)pyrene and ethylmethanesulfonate were purchased from Sigma and $CdCl_2$ from Merck. *Exposure.* Fibres were added to 0.8×10^6 cells/53 cm² flask during inoculation. EMS and $CdCl_2$ were added at the same time. One d later flasks were washed with PBS and 7.5 µg/ml BrdU added in new growth medium. Cultures were terminated 24-26 h later, i.e. after 2 cell cycles. BaP-exposure took place between the first and second halves of incubation; together with S9-mix (Fresenius, Bad Homburg) cells were shaken gently for 1 h at 37°C, washed and refed with medium containing BrdU. Preparation of slides and SCE stain. The methodology of Perry and Wolf (14) was employed.

RESULTS AND DISCUSSION

Exposure of B14F28 cells to chrysotile as well as crocidolite resulted in a high incidence of polyploid cells (Fig. 1). The production of SCE in several experiments is compiled in Figs. 2 and 3. It can be seen that chrysotile does not produce a spectacular effect, though it is significantly effective at concentrations exceeding 40 µg/ml (≙ 0.75 $\mu g/cm^2$). Crocidolite is a slightly better inducer of SCE.

Combined exposure of the fibres with the other mutagens yielded the following result: In three tests where cells were simultaneously treated with 150 $\mu g/10^6$ cells CFF or KFF and rising concentrations of EMS, there was no synergism detectable (Fig. 4). Combination of CFF and KFF with BaP also proved negative (Fig. 5), as did the combined treatment with asbestos and $CdCl_2$ (Fig. 6).

The present study demonstrates a weak but significant mutagenic effect of chrysotile and crocidolite in terms of SCE confirming the results of three previous working groups (2, 11, 12). The frequently documented fibre effects on ploidy were also highly evident in our cells (2, 5). However, none of the simultaneously applied other mutagens, though in themselves effective inducers of SCE, succeeded in enhancing SCE over and above the expected numbers: no synergism was noted.

The results should be discussed in the context of experimental investigations showing synergistic effects. Treatment of C3H mouse cells with crocidolite or amosite in combination with γ-rays produced no elevated SCE with the fibres alone, but a potentiated response together with the γ-rays. In addition, radiation enhanced the morphological transformation of the cells in a more than additive way (9). A synergism between BaP and asbestos was noted with respect to the morphological transformation of Syrian hamster embryo cells (15). *In vivo* an interaction between a nitrosamine, Cd and crocidolite was obtained. The lung tumour rate of rats receiving intratracheal instillation of all 3 components showed an apparent synergism, though on the other hand the combination of Cd and crocidolite alone caused no increase in bronchial carcinomas (16).

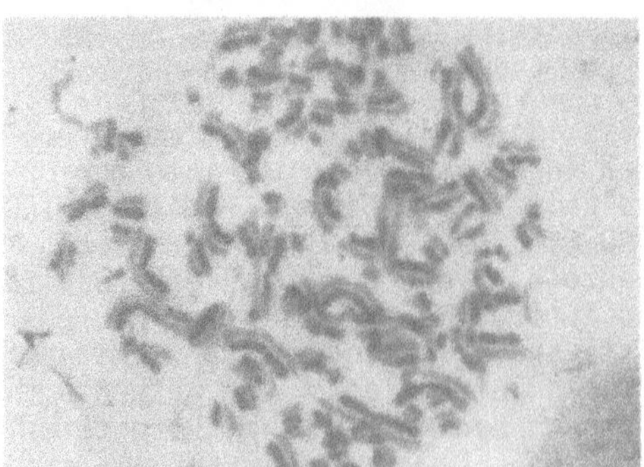

Fig. 1. Hexaploid cell induced by chrysotile

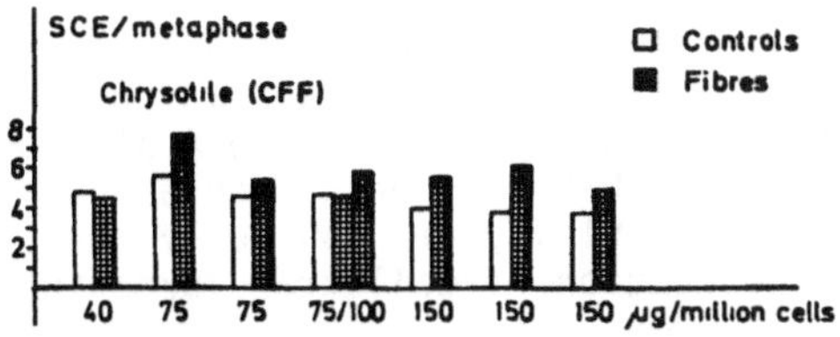

Fig. 2. SCE induction by different concentrations of chrysotile

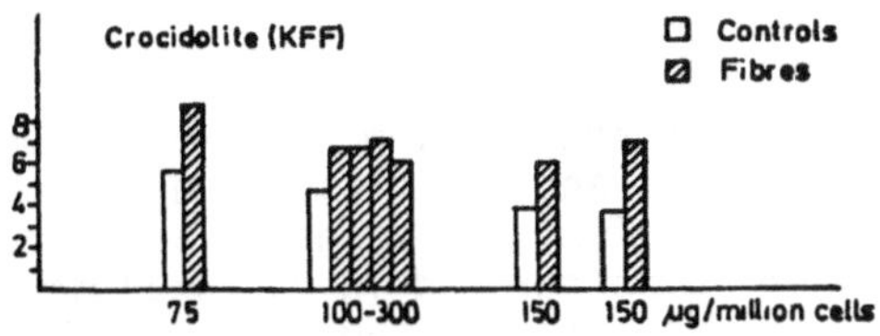

Fig. 3. SCE induction by different concentrations of crocidolite

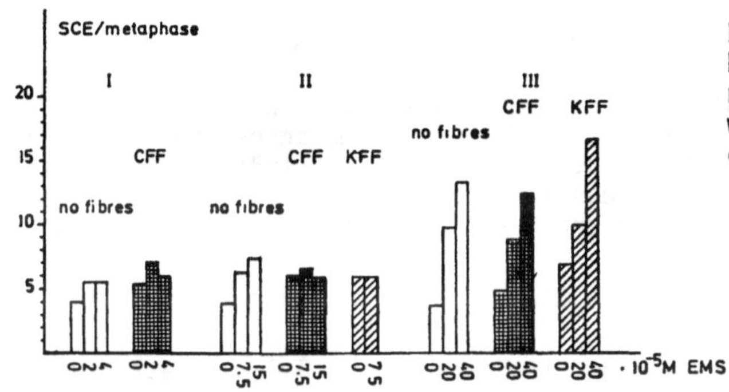

Fig. 4. SCE induction by ethylmethanesulfonate in combination with chrysotile and crocidolite

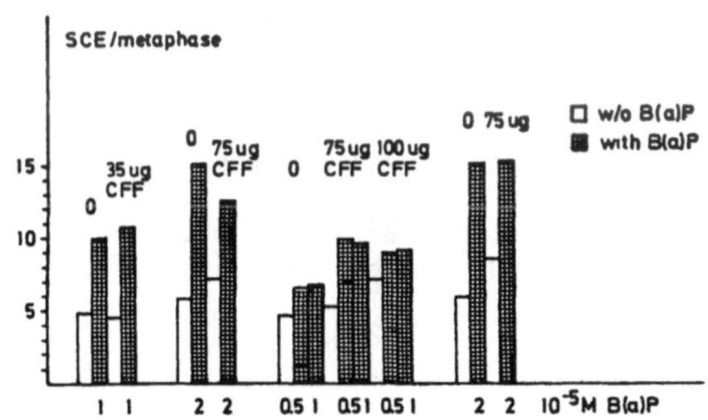

Fig. 5. SCE induction by benzo(a)pyrene in combination with chrysotile and crocidolite

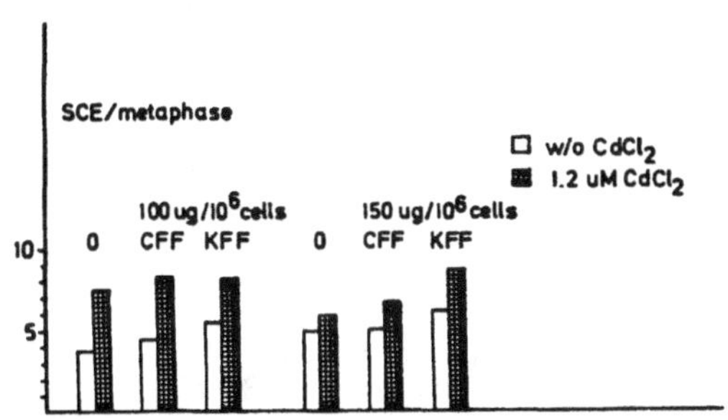

Fig.6. SCE induction by cadmium chloride in combination with chrysotile and crocidolite

In addition to the fact that the differing results were obtained using different test methods in vivo and in vitro, discrepancies may be explained by the finding that the induction of SCE is not the most important genotoxic effect of asbestos fibres. Other effects such as chromosome aberrations including alterations of ploidy are obviously more significant, possibly also for the production of cell transformation and cancer induction.

SUMMARY AND CONCLUSIONS

Synergistic interactions between asbestos and smoking have been documented in epidemiology and between asbestos and benzo(a)pyrene for cell transformation in vitro. Other genotoxic effects of asbestos are chromosome aberrations including aneuploidy demonstrated in cultured mammalian cells, while results concerning point mutation and the induction of sister chromatid exchanges are still controversial.

In the present study chrysotile and crocidolite produced weak but significant increases in SCE, while ethylmethanesulfonate, benzo(a)pyrene and cadmium chloride alone caused clear dose-dependent elevations. The combination of the fibres with the three other mutagens did not result in greater than additive SCE levels. It can be concluded that the genetic events producing SCE are less important than other genotoxic effects for the induction of heritable changes leading to the appearance of cancer cells.

REFERENCES

1. Oshimura, M., T.W. Hesterberg, T. Tsutsui, and J.C. Barrett (1984) Cancer Res., 44, 5017 - 5022.
2. Kelsey, K.T., E. Yano, H.L. Liber, and J.B. Little (1986) Br. J. Cancer, 54, 107 - 114.
3. Kenne, K., S. Ljungquist, and N.R. Ringertz (1986) Environ. Res., 39, 448 - 464.
4. Lavappa, K.S., M.M. Fu, and S.S. Epstein (1975) Environ. Res., 10, 165 - 173.
5. Sincock, A. and M. Seabright (1975) Nature, 257, 56 - 58.
6. Huang, S.L., D. Saggioro, H. Michelmann, and H.V. Malling (1978) Mutation Res., 57, 225 - 232.
7. Huang, S.L. (1979) Mutation Res. , 68, 265 - 274.
8. Casey, G. (1983) Mutation Res., 116, 369 - 377.
9. Hei, T.K., C.R. Geard, R.S. Osmak, and M. Travisano (1985) Br. J. Cancer, 52, 591 - 597.
10. Price-Jones, M.J., G. Gubbings, and M. Chamberlain (1980) Mutation Res., 79, 331 - 336.
11. Livingston, G.K., W.N. Rom, and M.V. Morris (1980) J. Environ. Pathol. Toxicol., 4, 373 - 382.
12. Babu, K.A., S.K. Nigam, B.C. Lakkad, D.K. Bhatt, A.B. Karnik, K.N. Thakore, S.K. Kashyap, and S.K. Chatterjee (1981) Environ. Res., 24, 325 - 329.
13. Takenaka, S., H. Oldiges, H. König, D. Hochrainer, and G. Oberdörster (1983) J. Nat. Cancer Inst., 70, 373 - 376.
14. Perry, P. and S. Wolf (1974) Nature, 251, 156 - 158.
15. Di Paolo, J.A., A.J. De Marinis, and J. Doniger (1983) Pharmacology, 27, 65 - 73.
16. Harrison, P.T.C. and J.C. Heath (1986) Carcinogenesis, 7, 1903-1908.

EVALUATION OF GENOTOXIC EFFECTS BY ENVIRONMENTAL POLLUTANTS USING CELL CULTURE SYSTEMS

Hadnagy, W. and N.H. Seemayer
Medical Institute of Environmental Hygiene at the University of Düsseldorf, Gurlittstr. 53, D-4000 Düsseldorf, FRG

INTRODUCTION

In a situation of growing environmental pollution, human population is exposed to a large number of chemical substances, which are extremely complex in nature. In the ambient air many known mutagens and carcinogens have been detected, which can lead to health hazards by respiration (1,2). According to epidemiological studies, there is evidence that air pollution may be a contributory factor to the increased risk for lung cancer among human population (3, 4). For this reason, early and rapid detection of the mutagenic and carcinogenic potential of harmful substances occurring in the atmosphere is of great importance (5). Since most mutagens and carcinogens produce cytologically visible damage to chromosomes(6), cytogenetic short term tests using human and rodent tissue culture cells seems to be of considerable value for predicting a genotoxic potential of environmental pollutants. On this base we have carried out a comparative study with a particulate extract of city smog and of automobile exhaust, respectively.

MATERIAL AND METHODS

City smog (No. 32) was sampled in 1985 in the highly industrialized Rhine-Ruhr area on fiberglass filters with a high volume sampler (HSV 150). Collected particles were extracted with dichloromethane. For genotoxic testing, the extract was quantitatively transferred to dimethyl sulphoxide (DMSO). Dosage was calculated by the benzo(a)pyrene content from which corresponding air volumes were determined.

Automobile exhaust (No. 1) was generated by a gasoline-powered engine run on the U.S. driving cycle (FTP-75). The exhaust was immediately diluted and particles were collected on a PVC-filter. Extraction was carried out under reflux using DMSO.

Test methods employed were analyses of chromosomal aberrations and sister chromatid exchanges (SCEs), determination of cell cycle kinetics, evaluation of mitotic index, recording of abnormal cell division stages and determination of numerical chromosome alterations as aneuploidy and polyploidy. Test systems used were human peripheral lymphocytes and Chinese hamster lung cells line V79. Treatment of the cell cultures with test substances and preparation techniques are described elsewhere (7, 8, 9).

RESULTS

Tab. 1 shows the results of the chromosome analysis in human peripheral lymphocytes after treatment with city smog extract. According to the tested concentrations corresponding to an air volume of 1.87 and 3.74 cbm, city smog extract leads to a significant increase of chromosomal aberrations (F-test; $p < 0.05$) in a dose-dependent manner.

Table 1. Frequency of chromosomal aberrations (without gaps) in human lymphocytes treated with city smog extract

Air volume cbm	Bap-Equ. ng/ml	Cells scored	Chromosomal aberrations per 100 cells ± S.E.
DMSO-control	-	300	1.6 ± 0.7
1.87	2.05	300	5.3 ± 1.3
3.75	4.10	300	6.7 ± 1.5

The results of the evaluation of SCEs are presented on Tab. 2 showing a distinct dose-dependent increase of SCE-frequencies. Additionally, as an indicator of cytotoxicity, the distribution of 1st, 2nd and 3rd division metaphases are taken into account revealing a pronounced cell cycle delay with dosage as shown by the given replication index (RI).

Table 2. SCE frequency and replication index (RI) after treatment of human lymphocytes with city smog extract

Air volume cbm	Bap-Equ. ng/ml	SCEs/metaphase ± S.E.	RI
DMSO-control	-	11.4 ± 0.6	1.82
0.06	0.06	11.8 ± 0.6	1.86
0.12	0.13	12.5 ± 0.6	1.80
0.23	0.24	14.1 ± 0.7	1.80
0.47	0.51	14.7 ± 0.7	1.60
0.94	1.03	15.2 ± 0.7	1.58
1.87	2.05	17.0 ± 0.8	1.59
3.74	4.10	18.8 ± 0.8	1.40

Employing the same cell system for testing the particulate extract from automobile exhaust (EPM), no induction of chromosomal aberrations could be detected. Even with the SCE-test, otherwise a very sensitive test for genotoxic substances, no considerable increase in the SCE-frequencies could be found (Tab. 3). Cell cycle delay by means of the replication index was only observed at the highest tested EPM-concentration.

Table 3. SCE frequency and replication index (RI) after treatment of human lymphocytes with EPM

EPM μg/ml	SCEs/metaphase ± S.E.	RI
DMSO-control	11.3 ± 0.6	1.64
0.08	11.2 ± 0.6	1.54
0.15	11.5 ± 0.6	1.58
0.31	10.7 ± 0.6	1.58
0.62	9.9 ± 0.6	1.55
1.24	11.7 ± 0.6	1.50
2.48	14.8 ± 0.7	1.55
4.95	12.9 ± 0.7	1.34

On the contrary, treatment of V79 cells with EPM revealed a pronounced increase of the mitotic index and the induction of abnormal cell division stages, which could exclusively be classified as so-called C-type mitoses (Tab. 4).

Table 4. Evaluation of mitotic index an C-type mitoses after treatment of V79 cells with EPM

EPM μg/ml	Mitotic index %	C-type mitoses %
DMSO-control	2.5	3
4.95	36.1	84
9.90	38.4	67

These effects could be attributed to disturbances of the spindle apparatus. As spindle disturbances may lead to numerical chromosome alterations, EPM-treated V79 cells were investigated for this cytogenetic endpoint. The results are presented in Fig. 1, showing an increase in the percentage of aneuploid and polyploid cells in a dose-dependent manner. As compared to colcemid, a known spindle poison, similar distribution patterns of diploid, aneuploid, and polyploid cells were obtained at EPM-concentrations of 4.95 and 9.90 μg/ml.

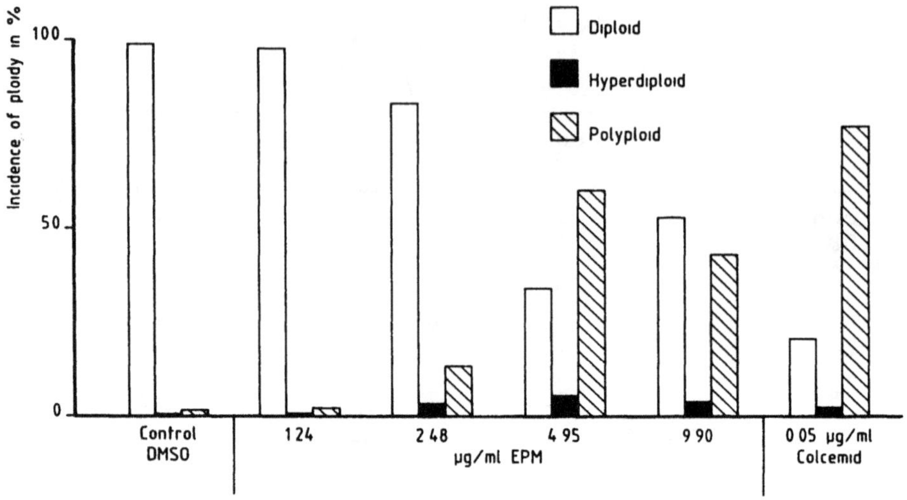

Figure 1. Induction of numerical chromosome alterations after treatment of V79 cells with EPM (published in ref. 7)

SUMMARY AND CONCLUSIONS

Results demonstrate that particulate extracts of city smog and automobile exhaust contain highly reactive substances being able to induce cytogenetic effects in human or rodent cell cultures. However, important differences between the two tested samples were observed. Whereas city smog extract induced direct effects on chromosomes as chromosomal aberrations and sister chromatid exchanges, particulate extract from automobile exhaust produced numerical chromosome alterations as a result of spindle disturbances. Since chromosomal aberrations and numerical chromosome alterations as aneuploidy and polyploidy may play a role in carcinogenesis (10), both chemical mixtures have to be considered as potentially carcinogenic. Based on data shown here and early reported (11), it is obvious that cytogenetic testing is of considerable value for predicting carcinogenic risk by environmental pollutants.

REFERENCES

(1) Helmes, C.T., D.L. Atkinson, J. Jaffer, C.C. Sigman, K.L. Thompson, M.I. Kelesey, H.F. Kraybill and J.I. Munn (1982) Evaluation and classification of the potential carcinogenicity of organic air pollutants, J.Environ.Sci. Health, A17, 321-389.

(2) Schlipköter, H.W. (1983) Lufthygienische Probleme der Großstadt, Arcus, 5, 244-250.

(3) Carnow, B.W. (1978) The urban factor and lung Cancer: Cigarette smoking or air pollution, Environ. Health Perspect., 22, 17-21.

(4) Doll, R. (1978) Atmospheric pollution and lung cancer, Environ. Health Perspect. 22, 23-31.

(5) Seemayer, N.H., W. Hadnagy and R. Tomingas (1987) Mutagenic and carcinogenic effects of airborne particulate matter from polluted areas on human tissue cultures, in: G.Boehm and R.M. Leuschner (Eds), Advances in aerobiology, Birkhäuser Verlag, Basel, pp. 231-234.

(6) Sorsa, M. (1980) Cytogenetic methods on the detection of chemical carcinogens, J. Toxicol.Environ.Health, 6, 1077-1080.

(7) Hadnagy, W. and N.H. Seemayer (1986) Induction of C-type metaphases and aneuploidy in cultures of V79 cells exposed to extract of automobile exhaust particulates, Mutagenesis 1, 445-448.

(8) Hadnagy, W., N.H. Seemayer and R. Tomingas (1986) Cytogenetic effects of airborne particulate matter in human lymphocytes in vitro, Mutat. Res., 175, 97-101.

(9) Seemayer, N.H., W.Hadnagy and R. Tomingas (1987) The effect of automobile exhaust particulates on cell viability, plating efficiency and cell division of mammalian tissue culture cells, Sci. Total Environ., 61, 107-115.

(10) Sandberg, A.A. (1983) A chromosomal hypothesis of oncogenesis, Cancer Genet. Cytogenet, 7, 277-285.

(11) Seemayer, N.H., W. Hadnagy, R.Tomingas and N.Manojlovic (1987) Cell- and genotoxic activities of airborne particulates from a highly industrialized region over a period of 11 years (1975-1986), J. Aerosol. Sci., 18, 721-724.

METABOLIC ACTIVATION AND CYTOTOXICITY OF AIRBORNE PARTICULATE MATTER

van Houdt J.J.
Department of Toxicology, Agricultural University, De Dreyen 12, 6703 BC Wageningen, The Netherlands.

INTRODUCTION

Mutagenic activity of airborne particulate matter is mainly located on the smallest particles.
The primary target cells for respirable particles are the macrophages bordering the lung alveoles. The regular clearance mechanism of the alveolar region, represented by the phagocytic activity of the macrophages may be impaired by the toxic action of respirable particles.
In this study the cytotoxicity of indoor and outdoor air samples on rat alveolar macrophages was determined.
Man is exposed to indirect acting mutagens for a far larger extend as to direct acting mutagens. Since lung, in particular the bronchial epithelium is the target tissue for carcinogenic substances it is not clear whether tests for mutagenicity with liver extracts might be reliable predictors of mutagenic activity in lung.
In this study also liver and lung homogenates of different species were compared towards their capacity of activating extracts of indoor and outdoor particulate matter.

MATERIALS AND METHODS

Preparation of aerosol extracts

Sampling of particulate matter was performed at indoor and outdoor locations, following a sampling and extraction technique, which is described previously (van Houdt et. al., 1987). Two samples were collected in livingrooms, one polluted by smoking and one polluted by wood combustion. Three samples were obtained outdoors: one in winter when wind direction was easterly and two in summer when wind direction was northerly. In summer a clean extract was obtained by sampling the same air mass two times.

Cytotoxicity testing

Alveolar macrophages were isolated from the lungs of female Wistar rats (±200 g) as described by Mason et. al.(1977). Alveolar macrophages were cultured following the methods described by Rietjens et. al. (1985). Cells were exposed by a two hour incubation to different amounts of aerosol extract. Phagocytosis was determined after incubation of the cells for 1,5 hour at 37°C in the presence of approximately 10^7 dead yeast cells, coloured by bioling them for 30 minutes in congo red PBS solution. All experiments were carried out on at least 5 occasions in duplicate.

Mutagenicity testing

The experiments were performed using the standard plate assay, according to Ames et. al.(1975), with minor modifications described pre-

viously (van Houdt, 1984).
Enzymic systems were prepared from livers and lungs of three month old male Wistar rats and six week old male Swiss mice. Half of the animals were pretreated with Aroclor 1254 (500mg/kg; Ames et.al., 1975).
Two independent tests were carried out in duplicate. In the results presented, the spontaneous revertants are substracted. Tester strain TA 98 was used in all experiments.
The level of metabolic activation of each enzyme preparate is expressed as metabolic factor, calculated from the slope of the dose(homogenate)-effect curve.

RESULTS

The results of the mutagenicity testing of 5 extracts, expressed as revertants per m^3, calculated from the slope of the linear part of the dose effect curve are presented in table 1. In the indoor samples only

Table 1: Mutagenic activity (rev/m^3) of indoor and oudoor extracts.

extract	-S9	+S9 (A1254) liver
outdoor clean	0	0
outdoor summer	0	0
outdoor winter	6	12
indoor woodcombustion	3	69
indoor smoking	2	98

a slight direct activity is detectable, while enzyme mediated mutagenic activity was strong. To a lesser extend this was also found in the outdoor sample, collected in winter. Both with and without liver S9 the samples, collected in summer hardly showed any mutagenicity.
The data presented in table 2 clearly show that extracts of airborne particulate matter may cause a dose dependent decrease in phagocytosis of rat alveolar macrophages.
Reduction of phagocytosis by extracts, obtained from different locations gave the following sequence of declining phagocytosis: indoor air polluted by woodcombustion, indoor air polluted by smoking, outdoor air collected in winter and outdoor air collected in summer. Clean air, just like DMSO did not reduce phagocytic activity.
Metabolic activation capacity of liver and lung homogenates towards extracts of airborne particulate matter is presented in table 3. Activating capacity was only determined in extracts, which show mutagenic activity in table 1.
Although activating capacity of the lung homogenates is less than the Aroclor induced liver homogenates, in all dose effect curves a high correlation ($p<0.05$) was found between the amount of added tissue and the number of revertants. Aroclor 1254 induction gives a strong increase of the metabolic activation capacity of mouse liver and rat liver. Activation by lung homogenates are comparable with uninduced liver homogenates. In contrast to indoor samples a relatively strong activation of outdoor samples was found by the rat tissue homogenates.
Extracts of indoor and outdoor particulate matter are activated at a comparative level by mouse tissue homogenates

Table 2: Effects of extracts of airborne particulate matter, collected at different locations on phagocytic activity of rat alveolar macrophages.

extract	m^3 of air	% phagocytosis
outdoor clean	20.0	96.9
	30.0	97.6
	40.0	97.1
	50.0	97.2
	60.0	97.4
outdoor summer	20.0	79.5
	30.0	43.8
	40.0	36.8
	50.0	29.1
	60.0	25.4
indoor woodcombustion	0.8	88.4
	1.2	83.7
	1.6	57.0
	2.0	30.0
	2.4	8.9

extract	m^3 of air	% phagocytosis
outdoor winter	6.0	93.1
	8.4	67.3
	9.6	48.8
	10.8	42.1
	12.0	32.4
	13.2	10.2
	15.0	4.6
indoor smoking	1.0	94.9
	1.5	89.1
	2.0	77.4
	2.5	64.4
	3.0	35.6
	3.5	27.0

Table 3: Metabolic activation capacity of liver and lung homogenates towards extraxts of indoor and outdoor particles.

	ind/woodcomb.	ind/smoking	outdoor
rat liver	3.2	3.2	17.4
rat liver Aroclor 1254	35.7	44.0	40.6
rat lung	2.4	1.4	5.6
rat lung Aroclor 1254	2.9	2.4	11.5
mouse liver	3.5	6.4	5.2
mouse liver Aroclor 1254	18.8	14.4	16.8
mouse lung	2.8	3.4	3.3
mouse lung Aroclor 1254	1.7	2.9	1.8

SUMMARY AND CONCLUSIONS

According to the results of previous studies (van Houdt et. al., 1984, 1986, 1987)mutagenic activity of the 5 extracts, determined in this study may be characterized as highly mutagenic indoor samples, a moderately mutagenic outdoor sample and two non-mutagenic outdoor samples. As macrophages show biochemical activity, which may be induced by environmental pollutants (Gardner, 1984) metabolic activation of premuta-

gens, bound at the surface of airborne particles may take place in macrophages (Cantrell, 1983; Harris et. al., 1978) Cytotoxicity of pollutants on macrophages thus may lead to the release of direct acting metabolites in the alveolar sacs.
The preliminary target cells for inhaled particles are the macrophages. Cultures of alveolar macrophages are frequently used for in vitro cytotoxicity studies, because they can be isolated from the lungs in a relatively easy way. Although in vivo macrophages are exposed to the particulate matter itself, the test, performed in this study may give the cytotoxic potency of the particles. Our results show that after addition of extracts of particulate matter a wide range of cytotoxicity on rat alveolar macrophages was observed. Clean air didn't cause a reduction of phagocytosis, polluted outdoor air gave a moderate action, while only 2 m^3 of indoor air gave a 50% reduction of phagocytosis.
In this study also the mutagenicity of 3 air samples was verified by the Salmonella/microsome assay, using different liver and lung homogenates. It is clear that Aroclor induced liver homogenates give the strongest activation capacity, in contrast to other homogenates, which mostly give a slight effect.Aroclor induction doesn't give a clear increase in activation capacity of lung homogenates. Comparison of the uninduced lung and liver homogenates shows that only minor differences in activation capacity are found. In contrast to mouse homogenates a rather strong activation of outdoor air was observed after adding rat homogenates. However, for all homogenates a positive relation ($p<0.05$) between amount of added tissue and number of revertants was found. Based on these results it may be concluded that the respiratory tract may be an important site for in vivo bioactivation of respirable particles.

REFERENCES

(1) Ames, B.N., J McCan and E. Yamasaki (1975) Methods for detecting carcinogens and mutagens with the Salmonella/mammalian microsome mutagenicity test. Mutation Res. 31, 347-364.
(2) Centrell, E.T., G.A. Warr, D.L. Busbee and R.R. Martin (1973) Induction of aryl hydrocarbon hydroxylase in human pulmonary alveolar macrophages by cigarette smoking. J. Air Invest. 52, 1884-1991.
(3) Gardner, D.E. (1984) Alterations in macrophage functions by environmental chemicals. Environ. Health Persp. 55, 343-358.
(4) Harris, C.C., I.C. Hsu, G.D. Stoner, B.T. Trump and J.K. Selkirk (1978) Human pulmonary macrophages metabolise benzo(a)pyrene to proximate and ultimate mutagens. Nature 272, 633-634.
(5) Houdt van J.J., W.M.F. Jongen, G.M. Aling and J.S.M. Boley (1984) Mutagenic activity of airborne particles inside and outside homes. Environ. Mutagen. 6, 861-869.
(6) Houdt van J.J., C.M.J. Daenen, J.S.M. Boley and G.M. Alink (1986) Contribution of wood stoves and open fire places to mutagenic activity of airborne particulate matter inside homes. Mutat. Res. 171, 91-98.
(7) Houdt van J.J., G.M. Alink and J.S.M. Boley (1987) Mutagenicity of airborne particles related to meteorological and air pollution parameters. The Sci. Total Environ. 61, 23-36.
(8) Mason, R.J., M.C. Williams, R.D. Greenleaf and J.A. Clements (1977) Isolation and properties of Type II alveolar cells from rat lung. Am. Rev. Resp. Dis. 115, 1015-1026.
(9) Rietjens I.M.C.M.; L. v Bree, M.Marra, M.C.M. Poelen, P.J.A. Rombout and G.M. Alink (1985) Gluthation pathway enzyme activities and the ozone sensitivity of lung cell populations derived from ozone exposed rats. Toxicology 37, 205-214.

MUTAGENS IN SURFACE AND WASTE WATER

Harald Müllerschön, H. G. Miltenburger
Laboratorium für Mutagenitätsprüfung (LMP), Technische Hochschule Darmstadt, D-6100 Darmstadt, F.R.G.

INTRODUCTION

In the last years it became evident that surface and waste waters contain a lot of mutagenic contaminants which can be detected and identified with suitable methods (1, 4, 5, 6, 8, 10). In a collaborative research project with the Bundesgesundheitsamt, Institut für Wasser-, Boden- und Lufthygiene, Berlin (sampling, extraction, and chemical analysis), and the Gesellschaft für Strahlen- und Umweltforschung, Institut für Toxikologie, Munich (Ames-test) native water samples and extracts were also tested for mutagenic effects in the cell line V79. It was the aim of the study to develop a combined method for the detection and valuation of mutagenicity in water samples. Mutagenic compounds could be detected in water from various sources in a wide range of concentrations in extracts and native water samples.

MATERIAL AND METHODS

The samples were tested for induction of chromosome aberrations, (CA), SCE and nuclear fragmentations. Extracts were prepared by three methods: 1. liquid-solid extraction by adsorption to a 1:1 mix of XAD-resins 2 and 7 (Röhm and Haas); 2. continuous liquid-liquid extraction (perforation) with diethyl-ether; 3. perforation with dichloromethane. The perforation technique has advantages from the analytical point of view but XAD is superior with regard to practical aspects e.g. it requires less time consuming laboratory procedures and lower amounts of highly purified solvents.

The V79 cells were taken from frozen stock cultures and thawed at least three days in advance of each experiment. They were seeded at a density of 1500 cells/cm² onto slides in 21 cm² Heraeus Quadriperm[R] chambers (Heraeus, D-6450 Hanau). The cells were cultivated at 37 °C in MEM (Hanks) supplied with 25 mM HEPES, 10 % fetal calf serum (FCS), 2.5 µg/ml Amphotericin B, and 100 µg/ml Neomycin sulphate in a controlled atmosphere of 4 % CO_2.

Each two parallel cultures were treated with at least six concentrations of the test sample together with a medium control, a solvent control (usually 50 µl DMSO = 1 %; Merck, D-6100 Darmstadt), and a positive control (usually 10 mM ethyl methanesulphonate; Merck). 48 h after seeding the cells were treated in medium without FCS for a period of 4 h, rinsed once and resupplied with medium containing all additives.

The cell preparation was done 24 h after the start of the treatment. Colcemid[R] (Fluka, D-7910 Neu Ulm) was added to the medium at a concentration of 0.5 µg/ml to induce metaphase arrest 2.5 h prior to preparation. After 20 min. of incubation with a hypotonic 0.0375 M potassium chloride solution the cells were fixed with a 3:1 v/v mixture of methanol and acetic acid. Spreading of the metaphases on the slides was obtained by removing the slides from the fixative.

The slides were air dried and stained with a 0.4 % solution of acetic orcein (Merck). At least three concentrations with a mitotic index higher than 25 % as compared to the solvent control were evaluated microscopically for chromosome aberrations, SCE, or nuclear fragmentations. From each culture 50 (CA), 25 (SCE), or 1000 cells (nuc. frag.) were evaluated, respectively. A sample was classified mutagenic if a dose dependent threefold increase (SCE: doubling) of the mutation frequency could be observed.

RESULTS

Water samples were taken from various sources: rivers and tributaries; industrial waste water of different origin; water from municipal waste water plants of different size. With the exception of a few water samples where cytotoxic compounds inhibited cell growth mutagenicity could be detected in samples of all origins depending on the degree of preconcentration. In heavily polluted water samples of different industrial sources mutagenic activity could be detected in the native water sample without further concentration.

Fig. 1 shows the reproducibility of the extraction techniques used. A water sample from an industrial waste water plant was separated into six portions and extracted independently three times using the XAD-adsorption technique (samples I - III) (1, 5, 6, 9, 12) and three times the dichloromethane perforation technique (IV - VI).

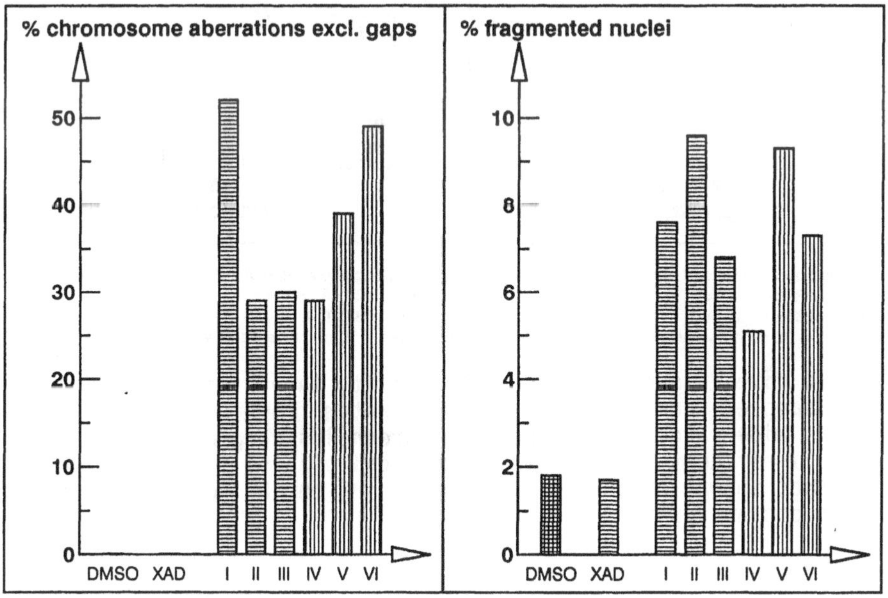

Fig. 1 Reproducibility of the extraction methods
DMSO = solvent control, 1 % solvent in medium; XAD = Blank, dissolved in DMSO, 1 %; 12,5 µl extract/5 ml MEM; Sample M 350; I-III = XAD extraction; IV-VI = Dichloromethane perforation; 1 µl extract = 1 ml native sample

The results of all six extracts show comparable mutagenicity at the same concentration of 12.5 µl extract in 5 ml of medium. The variability of the chromosome aberration data shown is due to the relatively small sample size (100 metaphases).

Fig. 2 and Fig. 3 compare results of two samples from a municipal waste water plant (fig. 2) and from a chemical plant (fig. 3). Considering the different concentration factors the extract of the municipal waste water sample showed significant mutagenicity at a concentration resembling 40 ml of water but the extract of the industrial water sample was mutagenic from 5 - 10 ml of water onwards. As shown in these two examples S9-mix generally reduced the mutagenic activity of the water samples.

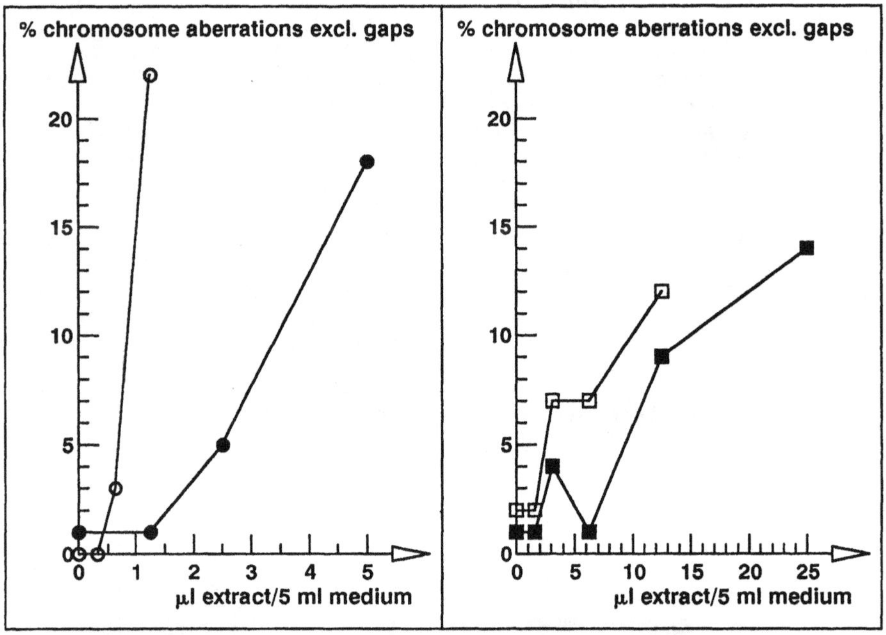

Fig. 2 Clastogenic effects of municipal waste water
—o— M 348 - S9-Mix
—●— M 348 + S9-Mix
1 µl extract = 30 ml native waste water

Fig. 3 Clastogenic effects of industrial waste water
—□— M 371 II - S9-Mix
—■— M 371 II + S9-Mix
1 µl extract = 1 ml native waste water

Fig. 4 gives a survey of the mutagenicity thresholds found so far in water samples from different sources. Water from industrial sources showed a wide range of mutagenicity thresholds ranging from 0.15 ml (pulp mill effluents) over 2.5 ml (chemical plants) to values above 100 ml (not shown) of native water depending on the concentration of mutagenic contaminants.

DISCUSSION

These results clearly demonstrate that there is a need of continuous observation and control to minimize the environmental contamination with mutagenic and potentially hazardous compounds. Further investigations are required to obtain a representative summary of the actual flux of mutagenic pollutants through the environment.

A combination of cytogenetic test systems, Ames-test, and chemical analysis allows to trace back mutagenic contaminations to their origins and to control the effectivity of technical improvements to reduce these contaminations. This method may thus be a helpful tool in reducing the environmental load with mutagenic substances.

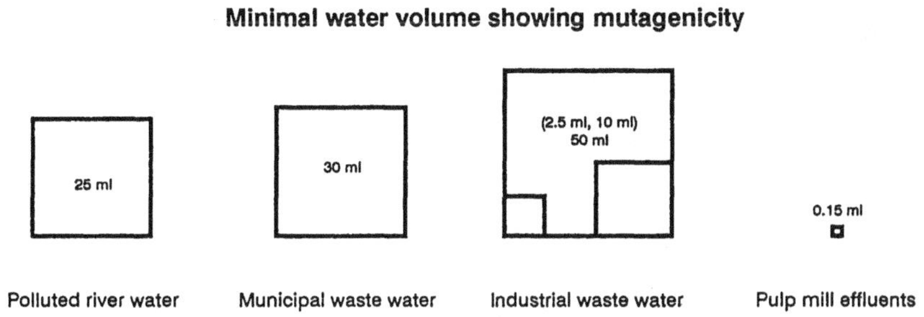

Fig. 4 Mutagenicity thresholds in water of different origin

Supported by the Bundesminister für Forschung und Technologie, D-5300 Bonn

REFERENCES

(1) Alfheim, I., A. Bjϕrseth , M. Mϕller, G.L. Fisher (1984) Characterization of microbial mutagens in complex samples - methodology and application. CRC Crit. Rev. Env. Control 14, Iss. 2, 91-150

(2) Ames, B.N., J. McCann, E. Yamasaki (1975) Methods for detecting carcinogens and mutagens with the Salmonella/mammalian microsome mutagenicity test. Mutation Res. 31, 347-464

(3) Bradley, M.O., B. Bhuyan, M.C. Francis, R. Langenbach, A. Peter son, E. Hubermann (1981) Mutagenesis by chemical agents in V79 Chinese hamster cells: A review and analysis of the literature. Mutation Res. 87, 81-142

(4) Douglas R., E.R. Nestmann, J.L. Betts, J.C. Mueller, E.G.-H. Lee, H.F. Stich, R.H.C. San, R.J.P Brouzes, A.L. Chmelauskas, H.D Paavila, C.C. Walden (1980) Mutagenic activity in pulp mill effluents. In: R.L. Jolley, R.B. Cumming, W.A. Bungs (Hrsg.) Proc Third Conf on Water Chlorination, Ann Arbor Science Publishers Inc., Ann Arbor, 865-879

(5) Epler, J.L. (1980) The use of short-term tests in the isolation and identification of chemical mutagens in complex mixtures. In: De Serres, F.J., A. Hollaender (Eds.) Chemical Mutagens, Principles and methods for their detection, Vol. 6, Plenum Press, New York, 239-270

(6) Grabow, W., J. Burgeru, C. Hilner (1981) Comparison of liquid-liquid extraction and resin adsorption for concentrating mutagens in Ames Salmonella/microsome assay on water. Bull. Environ. Contam. Toxicol. 27, 442 ff

(7) Ishidate, M., S. Odashima (1977) Chromosome tests with 134 compounds on Chinese hamster cells in vitro - a screening for chemical carcinogens. Mutation Res. 48, 337 351

(8) Jaylet, A., L. Gauthier, M. Fernandez (1987) Detection of mutagenicity in drinking water using a micronucleus test in newt larvae (Pleurodeles waltl). Mutagenesis 2, Nº 3, 211-220

(9) Junk, G., J.J. Richard, M.D. Grieser, D. Witiak, J.L. Witiak, M.D. Arguello, R. Vick, H.J. Svec, J.S. Fritz, G.V. Calder (1974) Use of macroreticular resins in the analysis of water for trace organic contaminations. J. Chromatog. 99, 745-762

(10) Kreijl, C.F. van, H.J Kool, M. de Vries, H.J. van Kranen, E. de Greef (1980) Mutagenic activity in the rivers Rhine and Meuse in the Netherlands. Sci. Total Environ. 15, 137-147

(11) Latt, S., J. Allen, S.E. Bloom, A. Carrano, E. Falke, D. Kram, E. Schneider, R. Schreck, R. Tice, B. Whitfield, S. Wolff (1981) Sister chromatid exchanges: a report of the GENE-TOX program. Mutation Res. 87, 17-62

(12) Rossum, P.v., R. Webb (1978) Isolation of organic water pollutants by XAD-resins and carbon, J. Chromatog. 150, 381-392

EFFECTS OF BENZENE AND OTHER LYSOSOMOTROPIC AGENTS ON ULTRAVIOLET LIGHT INDUCED PHOTOLYSIS OF LYSOSOMES IN HERPESVIRUS INFECTED CELLS

Orsi, E.V., S. Joshi, M. Bavlsik, M. Petersheim* and Y. Peng**
Biology and *Chemistry Departments, Seton Hall University, South Orange N.J. 07079, U.S.A. and **Virology Department, Wuhan University, Hupeh P.R.C.

INTRODUCTION

Benzene,although restricted in many countries for direct consumer utilization is still an important component in manufacturing and energy sources which finds its way into the environment. The interaction of this ubiquitous and mutagenic pollutant and common putative oncogenic viruses is of obvious concern. Host cell lysosome activity is known to modulate herpesvirus activity[1] and benzene which enhances cell transformation by HSV-2 virus[2] is known to destabilize lysosome integrity[3,4]. Hence,we decided to compare the effects of benzene and two lysosome stabilizers, hydrocortisone and chloroquine on lysosome membrane permeability during infection by two types of Herpes simplex virus HSV-1,type one and HSV-2, type two .

MATERIALS

Cell cultures: Mouse fibroblast line, 3T3-clone A 31, American Type Culture Collection (ATCC), Rockville, Md.
Culture Medium: MEM (Eagle,Earle's Salts) supplemented with 10% heat inactivated fetal bovine serum and standard antibiotics,GIBCO,Grand Id.NY.
Benzene: Thiophene-free grade, Fisher Scientific, Springfield, NJ. Used at a final concentration of 0.0024% (verified by gas chromatography).
Chloroquine: Sigma Chemicals, St. Louis,Mo.. Used at 0.020 nM/ml.
Hydrocortisone: Richter Chemicals, Milan. Used at 0.028 mg/ml.
Acridine Orange: J.T. Baker, Phillipsburg, NJ.. Used at 1:50,000.
Viruses: HSV-1 (MacIntyre strain),HSV-2 (G strain) both from A.T.T.C.

METHODS

Virus and Agent Interaction: In all cases subconfluent cell monolayers were inoculated with an MOI of about 2-5 PFU of virus and after 2 hours for attachment the inoculum was replaced with MEM + 5% serum or the same medium with benzene,hydrocortisone or chloroquine.
Total Virus Production: Monolayers formed in 24 well plates for each sampling period, provided 3 samples each of normal medium and compound exposed cells for both virus types. At appropriate intervals,the entire plate was frozen and thawed three times to release all cell associated virus and the entire contents of each well collected for assay.
Cell Associated and Free Virus: Monolayers were formed in 25cm^2 flasks. At 12 hours after infection, before any evidence of cell loss was seen in the normal medium monolayer, all flasks, three each from control and agent cultures were collected for both virus types. The overlying medium was collected and centrifuged to obtain cell-free medium and pellet. The pellet and the monolayer received 1.0 ml PBS (phosphate buffered saline) and frozen three times prior to assay. Two flasks, from all infected and uninfected groups were used to the cell number at the time of sampling.
Virus content of all samples was measured by the methylcellulose overlay method[5].
Lysosome Membrane Permeability: Membrane integrity was evaluated by the lysosome acridine orange retention method. This procedure has been described in detail[6]. In brief, cells were grown on 18 x 18 mm coverslips

placed in 35 mm petri dishes, infected and after attachment renewed with appropriate control media or those with the respective lysosomotropic agents. After 12 hours post infection incubation, acridine orange in MEM + 5% serum was added and reincubated for 30 minutes. The slips were then removed, the monolayer washed with PBS and the coverslip mounted with PBS and sealed with nail lacquer. Replicate fields of the same preparation were examined continuously with 410 nM illumination for two minutes and the amount of red fluorescence produced by the intralysosome acridine orange was measured by a photometer incorporated in the fluorescent compund microscope. At the start of the examination the photometer was adjusted to give an initial optical density of 0.30. The reduction in optical density as the acridine orange diffused from the lysosomes was recorded at half minute intervals. Figure 1 shows a typical plot where virus infected lysosomes retained acridine orange longer than those of uninfected cultures. With these plots dye retention half-life values were obtained much as they are in the case of radioactive decay. In this way lysosome membrane response to ultraviolet light was measured by the shift from red to green fluorescence as the acridine orange molecules diffused out of the lysosome and intralysosomal dye packing decreased.

In this study, three cultures for each group, i.e. uninfected normal medium, benzene, hydrocortisone and chloroquine were used along with their HSV-1 and HSV-2 infected counterparts. Addition of dye and microscopy were at staggered equal time intervals. In this way, time of incubation with dye as well as exposure to ultraviolet light were uniform with respect to the post-infection period.

Figure 1

Decreased Intralysosomal Fluorescence

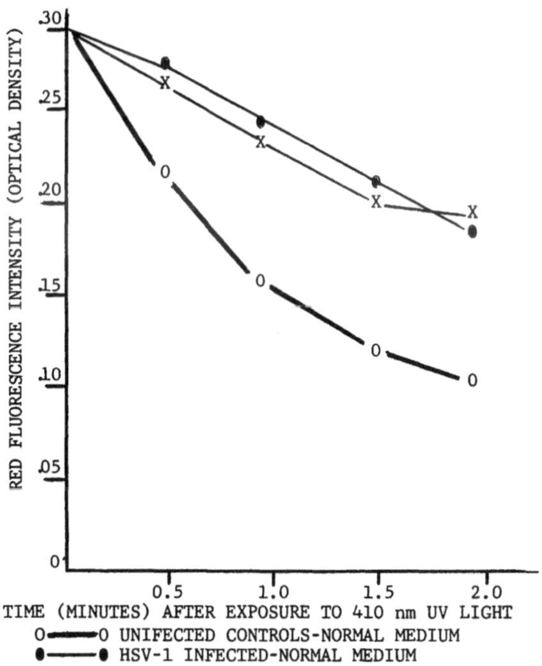

RESULTS

Throughout the course of infection (Table 1) HSV-1 and HSV-2 showed opposite responses to the various additives. At 12 hours, benzene enhanced HSV-2 replication but inhibited HSV-1. Hydrocortisone and chloroquine on the other hand markedly inhibited HSV-2. BY 36 hours, benzene produced almost a three fold increase of HSV-1 and inhibited HSV-2. Chloroquine never enhanced HSV-2 . Hydrocortisone, however stimulated replication of both types of Herpes simplex virus at the later stages of infection.

TABLE 1

RATIO TOTAL VIRUS WITH ADDITIVE MEDIUM/TOTAL VIRUS WITH NORMAL MEDIUM

	12 HOURS		18 HOURS		36 HOURS	
	HSV-1	HSV-2	HSV-1	HSV-2	HSV-1	HSV-2
BENZENE	0.82	1.30	1.45	0.45	2.80	0.30
HYDROCORTISONE	1.10	0.35	0.95	5.60	4.70	1.50
CHLOROQUINE	1.50	0.50	0.95	0.95	2.10	0.95

Another striking difference between the two virus types in responding to benzene is seen in Table 2. HSV-1 which normally releases virus more rapidly than type 2 showed a pronounced accumulation of virus but only if benzene was added to the medium.

TABLE 2

DISTRIBUTION OF HSV-1 AND HSV-2 TWELVE HOURS AFTER INFECTION

		PERCENTAGE OF TOTAL VIRUS			
		A	B	C	D
HSV-1	MONOLAYER	10.4	27.0	7.2	7.0
	PELLET	4.3	11.0	3.2	2.9
	RELEASED	85.3	62.0	89.6	90.1
	TOTAL PFU ($x 10^6$)	6.3	2.7	7.8	8.0
HSV-2	MONOLAYER	2.8	2.2	6.0	3.3
	PELLET	96.7	97.5	93.4	95.8
	RELEASED	0.5	0.3	0.6	0.9
	TOTAL PFU ($x 10^4$)	7.2	9.0	1.8	2.2

A = NORMAL MEDIUM
B = BENZENE
C = HYDROCORTISONE
D = CHLOROQUINE

As seen in Figure 1 and Table 3, infection with both types increased the retention of acridine orange in cultures with normal medium. All three compounds diminished the retention with HSV-2 . HSV-1 showed some loss of retention with hydrocortisone and chloroquine but benzene produced a marked enhancement of the ability of lysosomes to hold the dye after exposure to ultraviolet light.

TABLE 3

AVERAGE HALF LIFE (MINUTES) OF INTRALYSOSOME RED FLUORESCENCE

MEDIUM	UNINFECTED	HSV-1	HSV-2
NORMAL	1.0 ---	2.1 --	2.0 --
BENZENE	0.7 (0.7)	3.2(1.6)	1.6(0.8)
HYDROCORTISONE	0.5 (0.5)	1.7(0.9)	0.8(0.4)
CHLOROQUINE	0.8 (0.8)	1.2(0.6)	0.8(0.4)

( ) = RATIO ADDITIVE MEDIUM HALF LIFE/NORMAL MEDIUM HALF LIFE

SUMMARY AND CONCLUSIONS

HSV-1 and HSV-2 although closely related have often showed striking bio logic differences[7]. Here we report marked opposite responses to benzene hydrocortisone and chloroquine associated with modulation of virus replication and release and lysosome membrane photolysis by uv light. The enhanced retention of virus and intralysosome dye by benzene and HSV-1 is of particular interest since others have reported plasma membrane stabilization of HSV-1 infected HEp-2 cells but labilization of the membrane if infected by HSV-2 virus[8].

REFERENCES

(1) Smith, J. and E. DeHarven (1978) Herpes simplex virus and human Cyto megalovirus replication in Wi-38 cells. 3. Cytochemical localization in infected cells. J. Virol., 26 , 102-109.
(2) Johnson, E. (1982) Chemical interactions with Herpes simplex type 2 virus: Enhancement of transformation by selected chemical carcinogens and procarcinogens, Carcinogenesis,3(10), 1235-1240.
(3) Orsi, E.V., W. Zois and J. Howard (1981) Lysosome integrity and human herpesvirus infection in cell cultures, Fifth International Congress of Virology, Abstract P35/42, Strasbourg.
(4) Moszczynski, P. and J. Lisiewicz (1983) Effects of work environment contaminated with benzene, toluene and xylene on acid phosphotase activity in lymphocytes, Folia Haematol. Leipzig,110,948-954.
(5) Merchant, D.J., R.H. Kahn and W.H. Murphy,Jr. (1960) Handbook of cell and organ culture, Burgess Publishing Co., Minneapolis.
(6) Orsi, E.V., M. Bavlsik, R. Viera, M. Petersheim and O.F. Baturay (1986) Lysosome response and cytoskeleton alteration in cell cultures exposed to airborne lead. Third International Conference on Aerobiology, Basel , Experientia (in press).
(7) Nahamias, A. and W. Dowdle (1968) Antigenic and biologic differences in Herpesvirus hominis. In Progress in medical virology, Ed. J. Melnick, Karger Press, 10, 110-159.
(8) Hampl, H., J.R. Schlehofer and K.O. Habermehl (1981) Differences in the morphology of Herpes simplex virus infected cells II. Type specific membrane alterations of HSV-1 and HSV-2 infected cells, Med. Microbiol. - Immunol., 169, 209-223.

Supported in part by a grant to E.V. Orsi from the Research Grant Council of Seton Hall University.

TOXIC AND GENOTOXIC EFFECTS OF SO_2 OR NO_x, ALONE AND IN COMBINATION WITH CARCINOGENIC N-NITROSAMINES

B.L. Pool[1], R.G. Klein[1], S. Monarca[2], P. Schmezer[1], and W.J. Zeller[1]
[1]Institute for Toxicology and Chemotherapy, German Cancer Research Center, Im Neuenheimer Feld 280, D-6900 Heidelberg, F.R.G. and [2]Cettedra Di Igiene, Facolta di Farmacia, Universita di Perugia, 06100 Perugia, Italia

INTRODUCTION

The influence of common air pollutants SO_2 or NO_x on the genesis of tumors induced by chemical carcinogens has not yet been studied extensively. We may envisage the gases to act either inhibitory, additively or synergistically during the induction of tumors by carcinogens. Preliminary studies did show that lifelong SO_2 inhalation in combination with benzo(a)pyrene yielded an increased tumor rate in experimental animals (1). Studies on the interaction between SO_2 or NO_x with carcinogenic N-nitrosamines however, have not been reported. Therefore, we are performing a study on the interplay between SO_2 or NO_x and carcinogens at various levels, including in vitro studies, as well as a long term bioassay, which is still underway.

This report focusses on the determination of biochemical parameters, on in vitro studies using fetal hamster lung cells or rat hepatocytes, as well as short term in vivo studies. Various parameters of toxicity and genotoxicity were determined following exposure to SO_2 or NO_x. Combination effects were elucidated by additionally assaying various carcinogenic N-nitrosamines.

MATERIALS AND METHODS

Chemicals: N-Nitrosoacetoxymethylmethylamine (AMMN), N-nitrosodimethylamine (NDMA) and N-nitrosomethylbenzylamine (NMBeA) were kindly supplied by R. Preussmann and M. Wiessler of our institute. SO_2 and NO_x ($NO:NO_2$ = 1:1) were obtained from Messer Griesheim GmbH, Düsseldorf, FRG. Cell culture media, enzymes and reagents were commercially available (2).

Gaseous Exposure in vitro: A static exposure protocol was employed. Continuous analytical monitoring of gases proceeded in parallel (3).

Biochemical parameters: The activities of alkaline phosphatase (ALP), aspartate-amino-transferase (AST), alanine-amino-transferase (ALT), L-γ-glutamyltransferase (γ-GT) and lactate dehydrogenase (LDH) from a human serum standard (Precinorm U, Boehringer Mannheim, FRG) was determined after 1 to 5 hours gaseous exposure to SO_2 and NO_x using the enzyme kit, also from Boehringer.

Cells: Fetal hamster lung cells (FHLC) were isolated from fetuses of Syrian Golden hamsters on the 15th day of gestation (4). Hepatocytes were isolated by collagenase in situ perfusion (5) from adult Sprague Dawley rats which remained either untreated or received 50ppm SO_2 or NO_x for 2 weeks. 2×10^6 cells with a viability of > 80% were used for the analysis of toxic and genotoxic events.

Determination of toxic properties: The trypan blue exclusion assay was performed using an 0.4 % trypan blue solution and counting the stained and dye-excluding cells in a hemocytometer. Plating efficiency was determined by measuring colony forming units from treated cell cultures. The leakage of enzymes was determined by assaying the supernatant of cell suspensions using the Boehringer enzyme kit.

Determination of genotoxic properties: For the assay of induced DNA single strand breaks, suspensions of FHLC or rat hepatocytes were supplemented with 0.01 ml of DMSO or the nitrosamine dissolved in DMSO. The suspensions were exposed to the gases (air, SO_2, NO_x) for 1 hour. After determination of viability with the dye exclusion test the cells were counted, centrifuged and resuspended in buffer to be loaded on polycarbonate filters. Lysis of cells, washing procedures and fractionated DNA-elution was performed (2). For evaluation, the total amount of DNA retained on the filters of the treated cells was subtracted from the % DNA retained in the controls (C-T).

RESULTS AND DISCUSSION

Biochemical Parameters: The only influence either of the pollutant gases had on the native measurement of marker enzymes was seen in the reduction of LDH activity by SO_2. This reduction was due to HSO_3^- which inhibited enzyme activity to 31.2±3.1 , 17.8±2.1 , and 7.5±0.4 (% of control; n=5) at 0.1, 0.2, and 0.4 mMolar, respectively.

In vitro studies: Accordingly, when measuring liberation of enzymes into the media of treated cell suspensions, a decrease in LDH activity was observed instead of an increase which would indicate toxicity (Figure 1). This Figure also shows that the plating efficiency is distinctly decreased in the presence of SO_2 and marginally by NO_x In contrast, trypan blue exclusion was not a sensitive marker of toxicity.

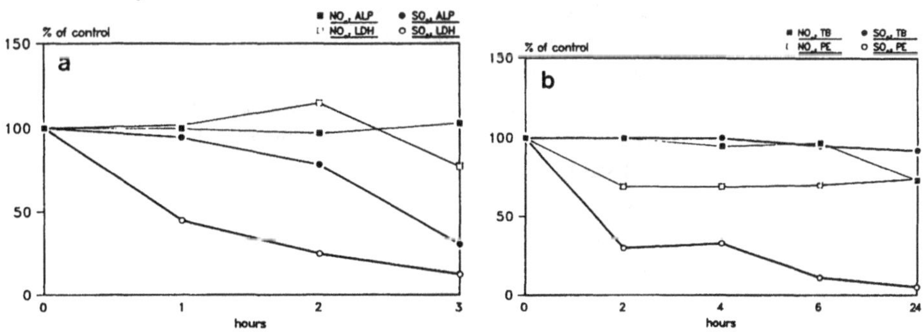

Figure 1: Determination of various toxicity parameters in vitro after exposing FHLC to SO_2 and NO_x; a. Enzyme leakage b. Trypanblue (TB) and plating efficiency (PE).

Figure 2 shows the effect of pollutants on the genotoxicity of AMMN in vitro. AMMN is a directly active derivative of NDMA and it is highly genotoxic. The rate of DNA single strand breaks, however, was lower in cells which simultaneously were exposed to SO_2 than those exposed to NO_x or those which remained untreated. The effects are marginal, but reproducible. Two major explanations may be given for this apparent protective effect by SO_2. One is that it is a result of conjugation of AMMN-intermediates with sulfate ions formed from sulfite within certain cells. Accordingly, similar protective effects were also found when adding $MgSO_4$ directly to the medium of hepatocyte suspensions. The second explanation is that, as a result of the inhibited LDH activity, an impaired cellular energy system may influence the rate of DNA single strand breakage by various mechanisms. This is supported by the findings of Maier et al.(6), who report a decrease in the level of ATP following treatment with sulfite.

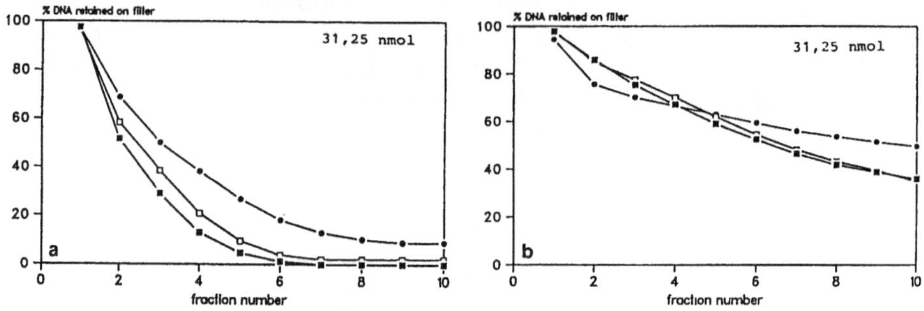

Figure 2: AMMN induction of DNA single strand breaks in FHLC (a) and rat hepatocytes (b). The cells remained unexposed (▪), or were additionally exposed to NO_x (□) or SO_2 (•).

Short term in vivo studies: For this we treat the animals with the pollutants, and measure the diverse parameters in explanted hepatocytes. After incubating them for 1 h with DMSO or the nitrosamines, enzyme determinations were performed in the medium. The most striking difference was found for the measurement of LDH activity in cell suspensions from SO_2 treated rats. In comparison to cell suspensions from untreated animals (148 U/L), only 1% activity was found. In these same cells we also determined the induction of DNA single strand breaks by various agents. Figure 3 shows there was also a reduced genotoxicity of AMMN, as well as of the 2 nitrosamines which must be metabolically activated (NDMA an NMBeA) in hepatocytes obtained from SO_2 treated rats. For these two compounds a third explanation of the inhibitory effects may be envisaged, and this is an impairment of their metabolic activation. This impairment is probably not due to a reduced presence of the responsible enzymes, since measurement of respective cytochrome P450 demethylase activities in microsomes from these hepatocytes showed that SO_2 even caused an induction (1.5 fold) of the enzyme (Monarca and Pasquini, pers. commun.). Therefore, the impairment is probably more due to a depletion of necessary cofactors and this again may be connected to the LDH inhibition.

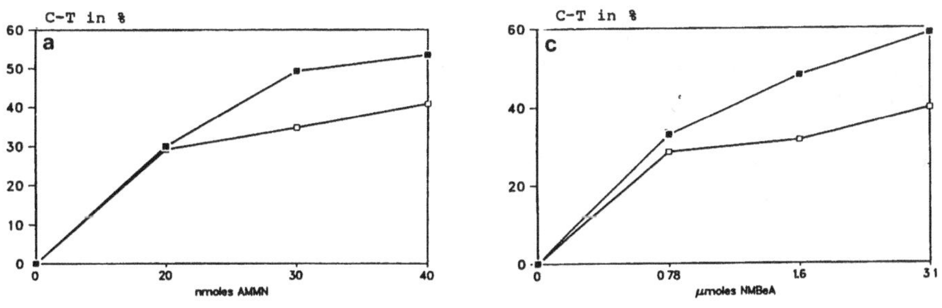

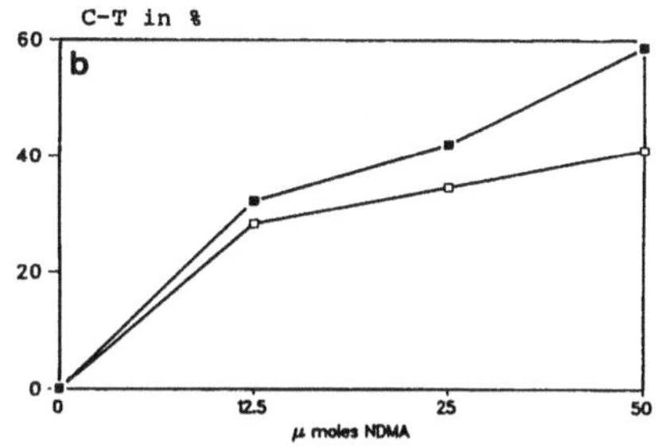

Figure 3: Genotoxic activity of 3 nitrosamines in hepatocytes derived from untreated rats (■) or from SO_2 exposed rats (□) a:AMMN; b:NDMA; c:NMBeA.

SUMMARY AND CONCLUSIONS

These results show measurable effects of SO_2, but not NO_x, on various parameters of toxicity and genotoxicity in vitro, as well as systemically after in vivo exposure to the gases. An important finding is the decreased activity of LDH following exposure to SO_2 or its unoxidized solution product SO_3^{2-}. This ultimately may be associated with a decreased level of ATP which in turn may explain why SO_2 causes a decreased plating efficiency of cells, or a reduction of genotoxicity, if this reduction is due to a reduced metabolic activation capacity of affected cells. Accordingly, alterations in the tumor profile of NDMA following SO_2 application may be a result. Definite modulating influences are presently being investigated in long term carcinogenicity assays combining inhalative applications of SO_2/NDMA and NO_x/NDMA to rats.

REFERENCES

(1) Henschler,D. (1984) Schwefeldioxid, in: Gesundheitsschädliche Arbeitsstoffe, toxikologisch-arbeitsmedizinische Begründung von MAK-Werten (Maximale Arbeitsplatz-Konzentration), DFG, Verlag Chemie, Weinheim, FRG.

(2) Pool,B.L. Eisenbrand,G. Preussmann,R. Schlehofer,J.R. Schmezer,P. Weber,H. and Wiessler,M. (1986) Detection of mutagens in bacteria and of DNA damage and amplified DNA sequences in mammalian cells as a systematic test strategy for elucidating biological activities of chemical carcinogens, Fd. Chem. Tox., 24, 685-691.

(3) Klein,R.G. Pool,B.L. Schmezer,P. and Zeller,W.J. In vitro exposure of mammalian cells to air pollutants, in preparation.

(4) Emura,M. Richter-Reichhelm,H.B. Schneider,P. and Mohr,U. (1980) Sensitivity of Syrian Golden hamster fetal lung cells to benzo(a)pyrene and other polycyclic hydrocarbons in vitro, Toxicology, 17, 149-155.

(5) Bradley,M.O. Dysart,G. Fitzsimmons,K. Harbach,R. Lewin,J. and Wolf,G. (1982) Measurements by filter elution of DNA single- and double-strand breaks in rat hepatocytes: effects of nitrosamines and γ-irradiation, Cancer Res., 42, 2592-2597.

(6) Maier,K. Beck,I. Hinze,H. Lenz,A.G. Leuschel,L. and Holzer,H. (1983) Biochemie der SO_2-Wirkung, in: Luftreinhaltung - Luftverschmutzung, AGF-Tagung, 3.-4. Nov., Bonn-Bad Godesberg, FRG, 60-62.

Acknowledgements: We are grateful to Mrs. S.Brendler, I.Janowsky, P.Klein and G.Vogt-Leucht for contributing to this work.

TARGET ORGAN CELLS AS IN VITRO TEST SYSTEMS FOR TOXICITY, MUTAGENICITY, DNA-REPAIR AND TRANSFORMATION BY ENVIRONMENTAL POLLUTANTS

Riebe, M., M. Emura, M. Aufderheide and U. Mohr
Institute for Experimental Pathology, Hannover Medical School, Konstanty-Gutschow-Str. 8, D-3000 Hannover 61, FRG

INTRODUCTION

Environmental pollutants do not influence or affect all tissues and cell types to the same extent. Organ specific reactions due to the particular metabolic capacities of each differentiated cell type are well known. This has to be considered when trying to define the hazardous potency of environmental chemicals. Since in vitro procedures today play an important role in toxicological research, type as well as origin of the cell cultures might be crucial points with respect to their possible applcation in risk assessment.
In the beginning in vitro researchers mainly used common fibroblastic cells which can relatively easily be isolated and cultivated to investigate the effects of toxicants. However, these fibroblastic cells have metabolic capacities different from those of epithelial cells. In long-term cultures external metabolizing systems have to be applied which are mostly derived from rat liver extracts. The metabolic pathways of the liver must, however, not be sigificant for other organs such as the lung. - The respiratory tract with its surface-covering epithelia of lung, bronchi and trachea is mainly exposed to environmental influences and is - as the primary site of contact - particularly at risk. This hypothesis is supported by the fact, that most respiratory tract tumors are of epithelial origin, i.e. carcinomas (1). Therefore it seemes to be a logical approach to use epithelial cells derived from the respiratory tract for investigations of problems specific to this organ system in vitro. However, the epithlial origin of in vitro grown cells is not the only requirement; the cells should, if they are to be used for toxicological research, at least exhibit their characteristic metabolism. Moreover, cells that are to be used in screening test systems for mutagens/carcinogens should be able to metabolize, i.e. activate or inactivate those chemicals.

MATERIALS AND METHODS

A clonal epithelial cell line (M3E3/C3) from the fetal Syrian golden hamster lung was used as target organ test system for cyto- and genotoxicity. The establishment of this cell line has previously been described (2).
Conventional Culture Conditions: Cells were grown on plastic surfaces with RPMI 1640 supplemented with 20% fetal bovine serum, glucose (3 mg/ml), sodium pyruvate (110 μg/ml), $FeSO_4.7H_2O$ (0,46 μg/ml), vitamin B12 (1,2 μg/ml), penicillin (100 units/ml), and streptomycin (100 μg/ml).
Complex Culture Conditions: Cells were grown on collagen gel substrate in a hormone-supplemented medium which contained 0,8 mg insulin, 0,18 mg Hydrocortisone sodium salt, 10 nmol 17β-estradiol, 2 μg epidermal growth factor, 2,5 mg human transferrin, 0,1 mg choleratoxin,

5,16 mg $CaCl_2$ and trace elements according to McKeehan et al. (3) in 98,1 ml of the growth medium described above. For further details on reagents, culture vessels, techniques etc. see Emura et al. (4).
Carcinogen Metabolism: Studies on the cellular carcinogen metabolism are described in detail by Jacob et al. (5).
Unscheduled DNA Synthesis (UDS): Cells were grown under conventional culture conditions on coverslips until confluency and were then treated with benzo(a)pyrene (BaP) or ethylnitrosourea (ENU) in the medium for 2h.(A detailed description of the ENU-incubation procedure is in preparation.) The medium was removed, the cells were washed twice with phosphate buffered saline (PBS) before they were incubated with a growth medium containing 2,5 μCi 3-H thymidin/ml for 3h. Autoradiographic analysis of UDS was done after removal of this "repair" medium, fixation of the cells, exposure to Kodak NTB 3 emulsion for 3 weeks and development of the specimens. Autoradiographic granules were counted per nucleus.
Mutations of the Hypoxanthine-Guanine-Phosphoribosyl-Transferase-Locus (HGPRT): Cells were grown under conventional culture conditions for 24 h and then exposed to 0,2 mg/ml ENU for 2h. Cells were trypsinized, replated and subcultured every other day for a period of 6 days to undergo phenotypic expression of 6-thioguanine resistance. Mutant frequency was determined after trypsination and replating of cells with 10 μg/ml of thioguanine in the medium.
Sister Chromatid Exchange (SCE): Cells were grown under conventional culture conditions for 24h and then treated with ENU for 2h. Further cultivation was performed in growth medium containing 10 μg/ml bromodeoxyuridine for two rounds of replication.During the last 3 hours of incubation 0,3 μg/ml demecolcin was added to the medium. Chromosome preparation and fluorescence plus Giemsa staining was done according to Speit et a. (6).
Chromosomal Aberrations: Cells were grown under conventional culture conditions up to 80% confluency. After 24 hours different amounts of ENU were added for 2h, then the cells were washed with fresh medium. After 2 days chromosome preparation was performed according to Rothfels and Siminovitch (7) after addition of 0,3 mg/ml demecolcine to the medium for the last 3h of incubation.

RESULTS AND DISCUSSION

The clonal epithelial cell line M3E3/C3 grows very rapid, undifferentiated and fibroblast-like under conventional culture conditions. However, the cell line has been characterized to be epithelial by means of various parameters. Keratin production was shown by immunological staining as already reported (8). Besides this the cell line can be induced to differentiate into certain types of epithelial cells which can be related to some corresponding cell types in vivo.
Basal cells in vivo might correspond to cells at an undifferentiated stage in vitro. By manipulation of the culture conditions, a metabolically stimulated cell type can be induced in vitro, which might correspond to the small mucus granule cell in vivo, a cell type that has been proposed to be the target cell of carcinogenic effects in the lung (9). Cells at this "step-1" differentiation are characterized by an increased content of endoplasmic reticulum and Golgi apparatus as well as small mucus granules. This indicates that at this predifferentiated level the cells are highly metabolically stimulated, a charac-

teristic that enables them to metabolize and activate indirect carcinogens very effectively. By further manipulation of the culture conditions a differentiation to mucus-producing cells can be induced (8). In the presented model, cells of this type are called "step-2" differentiated. - As a new characteristic of M3E3/C3 it should be mentioned here that the cell line has been shown to produce surfactant-like substances which would present clear evidence for alveolar type II cells (unpublished data). Therefore it is now believed that this cell line consists of multipotent bronchiolo-alveolar stem cells.
Under certain culture conditions, so-called complex conditions, the cell system is particularly susceptible to the effects of chemical carcinogens including polycyclic aromatic hydrocarbons. In vitro-transformation e.g. can only be induced in this certain stage of differentiation which is identical with the already mentioned "step-1" differentiation. The metabolic capacity of the cell line plays a key role with respect to their sensitivity against chemical carcinogens. As was shown in cooperation with Jacob et al. (5,10), the cell line M3E3/C3 metabolizes and activates polycyclic aromatic hydrocarbons very effectively. This represents a characteristic which has to be taken as an important prerequisit for a suitable in vitro test system. M3E3/C3, under complex culture conditions, does not only exhibit extremely high turnover rates, but also produces considerable amounts of crucial metabolites of chrysene (5). It has to be emphasized that those metabolism studies have been performed at passage numbers between 23 and 25 of M3E3/C3. I.e. the cell line does obviously not loose the potency to exhibit at least parts of the metabolic characteristics of lung epithelia. This certainly offers a number of possible applications of M3E3/C3 as an in vitro test model.
Besides cytotoxicity (4) benzo(a)pyrene or ethylnitrosourea induced UDS in M3E3/C3 as is shown in table 1. A number of further experiments have been conducted on the inducibility of sister chromatid exchanges (SCE), chromosomal aberrations and mutations of the HGPRT-locus. As a known potent genotoxic agent ethylnitrosourea was used. The results are presented in table 2.

Table 1: Induction of UDS in M3E3/C3 by chemical carcinogens (ENU, BaP).

Chemical	Dose (μg/ml)	Net grains/nucleus ± SD
ENU	0	5.4 ± 1.0
	100	19.5 ± 2.0
	200	34.4 ± 1.2
BaP	0	2.7 ± 1.2
	10	8.5 ± 2.1
	20	12.0 ± 2.5

Table 2: Induction of SCE, chromosomal aberrations and HGPRT-locus mutations in M3E3/C3 by ENU.

ENU-dose (μg/ml)	SCEs/metaphase ±SD	Aberrant metaphases (%)	HGPRT-locus mutations/10^5 cells ± SD
0	10.4 ± 2.8	0	0
200	32.2 ± 5.5	12	52.8 ± 6.2
400	40.0 ± 5.7	25	147.0 ± 13.4
800	-	37	95.4 ± 8.8

A number of studies presented here have been conducted using ENU as a model mutagen and carcinogen. ENU needs no metabolic activation by the treated cell, but regarding the before mentioned turnover rates of hydrocarbons as wll as the various studies performed so far using benzo(a)pyrene, it can certainly be presumed that the cell line will react similarly sensitive to other indirect carcinogens, at least to those belonging to the group of polycyclic aromatic hydrocarbons, which are particularly important with respect to air pollution and inhalation toxicology. - The final parameter investigated with this hamster cell line M3E3/C3 was in vitro transformation. As already mentioned before, only cells in a metabolically stimulated predifferentiated stage can be transformed, namely by benzo(a)pyrene as has been previously described in detail (4). Transformation is indicated by anchorage-independent growth of colonies in semi-solid medium. I.e. the assumed end-point of genozoxic events, which might be a transformed cell clone, can also be demonstrated by the cell system M3E3/C3. However, this is at least true for those in vitro described cells resembling the small mucus granule cells in vivo. The presented data would therefore possibly indicate that only relatively undifferentiated cells with an active capability for cell division could be target cells for chemical carcinogens. In contrast, terminally differentiated cells, in this system represented by bronchiolar mucus cells, are not division active cells and are thus not transformable.

SUMMARY AND CONCLUSIONS

A cell line which has been established from the fetal Syrian golden hamster lung was shown to differentiate into certain types of epithelial cells present in the respiratory tract in vivo by application of different culture conditions. The cell line M3E3/C3 is proposed as a target organ cell system for the testing of cyto- and genotoxic as well as transforming chemicals relevant in inhalation toxicology.

ACKNOWLEDGEMENTS

This study has been partly supported by a grant from DFVLR-PT Humanisierung des Arbeitslebens.

REFERENCES

(1) McDowell, E.M. and B.F. Trump (1983) Histogenesis of preneoplastic and neoplastic lesions in traceobronchial epithelium, Surv. Synth. Path. Res. 2, 235-279.

(2) Emura, M., H.-B. Richter-Reichhelm, W. Böning, R. Eichinger, C. Schoch, J. Althoff and U. Mohr (1982) A fetal respiratory epithelial cells line for studying some problems of transplacental carcinogenesis in Syrian golden hamsters, J. Cancer Res. Clin. Oncol. 104, 133-144

(3) McKeehan, W.L., K. A. McKeehan, S.L. Hammond and R.G. Ham (1977) Improved medium for clonal growth of human diploid fibroblasts at low concentrations of serum protein, In Vitro 13, 399-416

(4) Emura, M., U. Mohr, M. Riebe, M. Aufderheide and D.L. Dungworth (1987) Predisposition of cloned fetal hamster lung epithelial cells to transformation by a precarcinogen, benzo(a)pyrene, using growth hormone supplementation and collagen gel substratum, Cancer Res. 47, 1155-1160

(5) Jacob, J. G. Grimmer, G. Raab, M. Emura, M. Riebe and U. Mohr

(in press) Comparison of chrysene metabolism in epithelial human bronchial and Syrian hamster lung cells, Cancer Letters
(6) Speit,G. (1984) Considerations on the mechanism of differential staining of BrdU-substituted chromosomes, Hum. Genet. 67, 264-269
(7) Rothfels, K.H. and L. Siminovitch (1958) An air-drying technique for flattening chromosomes in mammalian cells grown in vitro, Stain Technol. 33, 73-77
(8) Emura, M., M. Riebe, M. Aufderheide and U. Mohr (1986) A fetal Syrian hamster lung cell line as an in vitro model for respiratory carcinogenesis, Exp. Pathol. 29, 211-216
(9) Trump, B.F., E.M. McDowell, F. Glavin, L.A. Barrett, P.J. Becci, W. Schürch, H.E. Kaiser and C.C. Harris (1978) The respiratory epithilium. III. Histogenesis of epidermoid metaplasia and carcinoma in situ in the human, J. Natl. Cancer Inst. 61, 563-575
(10) Grimmer, G. J. Jacob, A. Schmoldt, G. Raab, U. Mohr and M. Emura (1983) Metabolism of benzo(a)anthracene in hamster lung cells in culture in comparison to rat liver microsomes, Proceedings of 7th International Symposium of PAH, Columbus, Ohio on 28.10.1983

INDICATORS OF POTENTIAL HEALTH RISKS BY AIRBORNE PARTICULATES: CYTOTOXIC, MUTAGENIC AND CARCINOGENIC EFFECTS ON MAMMALIAN CELLS IN VITRO

Seemayer, N.H., W. Hadnagy, Heidrun Behrendt and R. Tomingas, Medical Institute of Environmental Hygiene at the University of Düsseldorf, Gurlittstr. 53, D-4000 Düsseldorf, F.R.G.

INTRODUCTION

Air pollution is a problem of growing medical concern. Traffic, urbanization and growing urbanization are the most important factors causing an increased contamination of the atmosphere with particulate and gaseous pollutants. Chronic inhalation of a great variety of air pollutants raises serious questions about effects on human health. Airborne pollutants, especially particulates contain more than 600 mostly organic substances among them recognized and potential mutagens and carcinogens (1, 2). According to Sawicki (3) air pollutants play an important role in cyto- and genotoxic processes f.i. in carcinogenesis, mutagenesis, teratogenesis and ageing. Airborne particulates with a diameter smaller than 5 μm represent respirable dust, which by inhalation reaches broncho-alveolar space in human lung affecting cell functions. The enormous alveolar surface of human lung of 150 m^2 is built up of an exceedingly thin and vulnerable tissue barrier between blood and air and can be easily damaged by air pollutants. The burden of air pollutants in highly contaminated regions may exceed lungs ability to eliminate and detoxify noxious substances. Therefore, long-term effects of living in a polluted atmosphere on human health must be considered (4). Three adverse effects induced by airborne particulates are of special medical importance: cytotoxicity, mutagenicity and carcinogenicity. For assessment of these effects cell cultures of human, primate and rodent origin offer special advantages as indicators of potential health risks to man (5, 6). In this study we present data on 26 samples of airborne particulates or city smog, which were collected between 1969 and 1986 in the heavily industrialized Rhine-Ruhr region (F.R.G.). All samples were extracted by organic solvents and analyzed for their cytotoxic and genotoxic potency utilizing rodent and human tissue culture cells

MATERIALS AND METHODS

Collection and extraction of samples of airborne particulates or city smog as well as cultivation of macrophages and determination of cell viability has been described previously (7, 8). Reduction of "plating efficiency" of mammalian cells in cultures, estimation of "sister chromatid exchanges" in human lymphocyte cultures and the "cell transformations assay" applying Syrian hamster kidney cultures infected with Simian virus (SV-) 40 were outlined in earlier papers (5, 9, 10, 11). Biochemical estimation of DNA synthesis and cell cycle analysis by autoradiography has been published previously (12 - 16). Pulse cytophometric cell cycle analysis of synchronized cells has been reported (17).

RESULTS AND DISCUSSION

Cytotoxicity of city smog

Most samples of city smog induced pronounced cytotoxic effects on mammalian cells in vitro affecting cell structures and diverse cell functions. In presence of extracts of city smog mouse macrophages of line IC-21 and human monocytes/macrophages revealed a dose dependent loss of cell viability (7, 8, 18). Morphological alterations of exposed macrophages (line IC-21) were characterized by "bleb" formation at the cell surface, vacuolization of cytoplasm, formation of multinucleated cells, pycnosis of cell nuclei and cytolysis (19, 20). Specific ultrastructural alterations of macrophages (line IC-21) were observed with extract and fractions of city smog (21). Phagocytosis of zymosan particles by human monocytes/macrophages in presence of extract of city smog or of exhaust particulates of a gasoline powered engine was remarkably reduced (unpublished observation). We have to emphasize that most sample of city smog led to a significant loss of cell viability (survival) already at concentrations of particulates equivalent to air volumes of 4-8 m^3. For comparison, human respiratory ventilation amounts to 12-14 m^3 per day in rest. Therefore, macrophages which play a pivotal role in the defense mechanism of the lung, could be affected by airborne particulates, especially under conditions of an "inversion" with substantially increased pollution.

Cell replications of mammalian cells was generally affected by extracts of city smog Cell growth of human diploid lung fibroblasts was strongly inhibited by extract and some fractions of city smog (7) as well as of Syrian hamster kidney cells and of kidney cells of Ceropithecus aethiops (12, 13). "Plating efficiency" of rodent and human cell lines (Syrian hamster kidney line 14-1b, Chinese hamster lung cell line V 79 and human type II pneumocyte line A 549) was strongly reduced by extracts of city smog after incubation periods of 24 and 120 h, respectively (5, 22). Cellular DNA synthesis and cell cycle progression of mammalian cells in vitro was altered in presence of extracts of city smog. Dose related inhibitions of DNA synthesis of Syrian hamster kidney and embryonic cells and of kidney cells of the primate Cercopithecus aethiops were demonstrated by autoradiography and biochemistry (12 - 16)). Profound perturbations of cell cycle progression of Syrian hamster kidney cells induced by extract and some fractions of city smog were detected by autoradiography employing the method of labelled mitosis (15, 16). Using the method of cell cycle analysis by pulse cytophotometry and synchronised Syrian hamster cells it was shown that cells in DNA synthesis are especially sensitive to the effect of city smog. A prolongation of the S-phase and a high accumulation of cells in G2-phase was observed (17).

Mutagenicity and carcinogenicity of city smog

Mutagenic activity of extrats of city smog was analyzed by induction of "sister chromatid exchanges" (SCE) in cultures of human lymphocytes, V 79 and A 549 cells. "Sister chromatid exchange" is a sensitive bioassay for detecting DNA damage by environmental chemicals.

We observed significant inductions of "sister chromatid exchanges" with quantities of extracts equivalent to particulates from less than 0.5 m^3 of air (5, 10, 11, 22). A comparable dose dependent induction of "sister chromatid exchanges" was obtained using the Chinese hamster lung cell line V 79 and human type II pneumocyte line A 549 (unpublished observations).

Furthermore, chromosomal aberrations occured in a dose dependent manner in human lymphocyte cultures accompanied by a pronounced cell cycle delay at higher concentrations of extract (10).

Induction of "point mutations" at the HGPRT-locus were observed with V 79 cells treated with extract (23). "Enhancement" of virus-induced cell transformation appears to be a very sensitive indicator for chemical agents with the potential for carcinogenicity and therefore may be a useful tool for detection of these substances in the environment, especially in the atmosphere (5, 24). Strong "enhancement" of malignant cell transformation occured in a linear dose response relationship with Syrian hamster kidney cultures pretreated with extracts of city smog and consecutively infected with Simian virus (SV-) 40 (6, 8, 9, 22, 25). It has to be stressed that already small amounts of extracts,

equivalent to particulates from 0.1 -0.5 m^3 of air led to a significant increase of cell transformation frequency (5, 22, 25, 26).
Genotoxic data of extracts of city smog (collected between 1975 and 1986) are summarized on fig. 1. Presented are the lowest quantities of extracted particulates expressed as air volumes of collection (in m^3) which led to significant inductions of "sister chromatid exchanges" and to "enhancement" of malignant cell transformation ($p < 0.001$).

SUMMARY AND CONCLUSIONS

We report about the cytotoxic and genotoxic potency of 26 samples of airborne particulates or city smog collected between 1969 and 1986 in the highly industrialized Rhine-Ruhr region. Extracts of samples were tested on human and rodent tissue culture cells.

Cytotoxicity of extracts induced a dose related
- loss of cell viability of mouse and human macrophages
- impairment of phagocytosis of mouse and human macrophages
- inhibition of cell growth of rodent, primate and human tissue culture cells
- inhibition of DNA synthesis and alterations of cycle progression
- reduction of "plating efficiency" of rodent and human cell lines

Genotoxicity of extracts produced dose dependently
- induction of "sister chromatid exchanges" in human lymphocytes cultures, in Chinese hamster cell line V 79 and human type II pneumocyte line A 549
- induction of "chromosomal aberrations" in human lymphocyte cultures
- induction of "point mutations" at the HGPRT-locus in V 79 cells
- strong "enhancement" of malignant cell transformation of SV40-infected Syrian hamster kidney cultures

Results demonstrate that rodent and human tissue culture cells can be used succesfully for evaluation of cytotoxic and genotoxic potency of airborne particulates. In the Rhine-Ruhr region concentration of airborne particulates declined since 1969 from 100-200 µg/m^3 of air to values of 30-50 µg/m^3 in 1986. Similarly, benzo(a)pyrene became reduced in the same period from 50-100 ng/m^3 to 10 ng/m^3 or less. Despite of this reduction, cytotoxic and genotoxic potency of airborne particulates estimated per volume of air in m^3 remained unchanged (22). We assume, that the high adsorptive capacity of airborne particulates for chemicals is responsible for this phenomenon. Data on cytotoxicity and genotoxicity suggest, that airborne particulates inhaled into the respiratory system provoke injuries on alveolar and bronchial tissue. Medical risk evaluation and body burden by airborne particulates must take into account a dayly respiratory ventilation in man of 12-14 m^3 in rest. Cytotoxicity shown by particulates from 4 m^3 of air must therefore be multiplied with a factor of 3-3,5, genotoxicity exerted by particulates from 0.1 m^3 with a factor of 120-140 to correspond to respiratory ventilation per day. This approach can be used to calculate relative body burden and potential health risk of airborne particulates in human beings.

City smog No	Place of collection	Period of collection
16	Duisburg	1975/76
17	Gelsenkirchen	1977
18	Duisburg	1979/80
19	Krahm	1978/79
20	Duisburg	1981
21	Krahm	1981
22	Dusseldorf	1982
23	Krahm	1982
24	Duisburg	1982
25	Krahm	1983
26	Duisburg	1983
27	Dusseldorf	1983
28	Krahm	1984
29	Duisburg	1984
30	Dusseldorf	1984
31	Dusseldorf	1985- 7
32	Dusseldorf	1985- 9
33	Dusseldorf	1985-12
34	Dusseldorf	1986- 1
35	Dusseldorf	1986-12

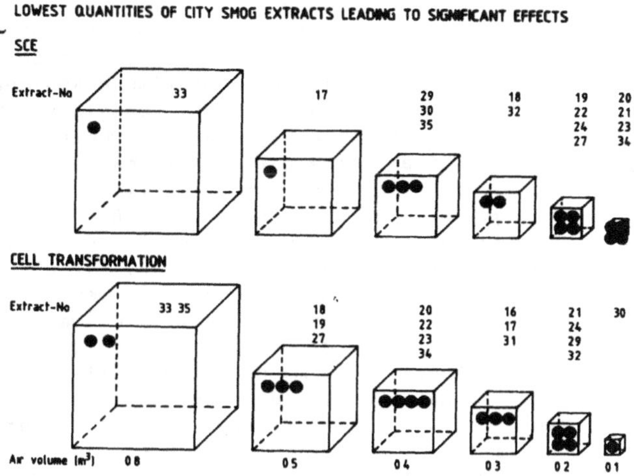

Fig. 1 Cubes of different sizes represent air volumes (0.8-0.1 m^3) related to quantities of airborne particulates showing mutagenic and carcinogenic activity

REFERENCES

(1) Helmes, C.T., D.L. Atkinson, J. Jaffer, C.C. Sigmann, K.L. Thompson, M.J. Kelsey, H.F. Kraybill, J.J. Munn (1982) Evaluation and classification of the potential carcinogenicity of organic air pollutants, J. Environ Sci. Health A17, 321-389.

(2) Schlipköter, H.-W. (1983) Lufthygienische Probleme der Großstadt, Arcus, 5, 244-255.

(3) Sawicki, E. (1977) Chemical composition and potential genotoxic aspects of polluted atmosphere. in: Air pollution and cancer in man. Ed. by U. Mohr, D. Schmähl and L. Tomatis, IARC, Lyon, pp. 127.

(4) Lave, L.B. and E.P. Seskin (1970) Air pollution and human health, Science 169, 723-733.

(5) Seemayer, N.H., N. Manojlovic, C.-C. Schürer and R. Tomingas (1984) Cell cultures as a tool for detection of cytotoxic, mutagenic and carcinogenic activity of airborne particulate matter, J. Aerosol. Sci 15, 426-430.

(6) Seemayer, N.H. (1978) Zelltransformation von Hamsternierenzellen in vitro unter dem Einfluß von atmosphärischen Feinstaubextrakten und dem Papovavirus SV40, Staub-Reinhalt. Luft 38, 254-258.

(7) Seemayer, N.H. and N. Manojlovic (1980) Cytotoxic effects of air pollutants on mammalian cells in vitro, Toxicology 17, 177-182.

(8) Seemayer, N.H., N. Manojlovic and A. Brockhaus (1976) The effect produced by fine-dust extracts on cells in vitro, with particular regard to carcinogenic components, Zbl. Bakt. Hyg., I. Abt. Orig. B 162, 27-33.

(9) Seemayer, N.H., N. Manojlovic und R. Tomingas (1980) Die Wirkung von kanzerogenen Stoffen, insbesondere von atmosphärischen Feinstaubextrakten, im Zelltransformationstest. In: Polyzyklische aromatische Kohlenwasserstoffe - VDI Berichte Nr. 358, VDI-Verlag, Düsseldorf, pp. 285-292.
(10) Hadnagy, W., N.H. Seemayer and R. Tomingas (1986) Cytogenetic effects of airborne particulate matter in human lymphocytes in vitro, Mutation Res. 175, 97-101.
(11) Hadnagy, W., and N.H. Seemayer (1987) Comparative investigation on the genotoxicity of city smog and automobile exhaust particulates, J. Aerosol Sci. 18, 697-699.
(12) de Ruiter, N., N. Manojlovic, N.H. Seemayer, G. Krampitz and H. Weisz (1978) Analysis of the biological effect of city smog extract. II. Effect of city smog extract on cell growth and DNA synthesis of hamster kidney cells in vitro, Zbl. Bakt. Hyg. I. Abt. Orig.B 166, 558-566.
(13) Manojlovic, N., N.H. Seemayer, N. de Ruiter, H. Weisz and S. Bauer (1978) Analysis of the biological effect of city smog extract. III. Comparative investigations on the effect of city smog extracts on cell replication and DNA synthesis of kidney cells in vitro from the primate Cercopithecus aethiops, Zbl. Bakt. Hyg. I. Abt. Orig. B 167, 90-97.
(14) Krampitz, G. and N.H. Seemayer (1980) Analysis of the biological effect of city smog extracts and their fractions on DNA synthesis of hamster kidney and embryonic cells and kidney cells of the primate Cercopithecus aethiops, Zbl. Bakt. Hyg. I. Abt. Orig. B 170, 1-16.
(15) Behmer, A. and N.H. Seemayer (1983) Biological effects of city smog VII. Strong alterations of the cell cycle and its phases in kidney cultures under the influence of extract and fractions of city smog collected in a heavily industrialized area, Zbl. Bakt. Hyg. I. Abt. Orig. B 178, 340-363.
(16) Seemayer, N.H. and G. Krampitz (1979) Autoradiographic investigations on the effect of city smog on DNA synthesis and cell cycle of mammalian cell in vitro II. Alterations of the cell cycle of hamster kidney cells and hamster embryonic cells in presence of city smog extract, Zbl. Bakt. Hyg. I. Abt. Orig. B 168, 403-421.
(17) Giebel, P. and N.H. Seemayer (1984) Biological effects of city smog. VIII. Impulscytophotometric analysis of the cell cycle of synchronized Syrian hamster kidney cultures (line 14-1b), Zbl. Bakt. Hyg. I.Abt. Orig. B 179, 406-430.
(18) Seemayer, N.H., N. de Ruiter, N. Manojlovic and H. Weisz (1977) Analysis of the biological effect of city smog extract. I. Cytotoxicity of city smog extract from urban area and of polycyclic aromatic hydrocarbons on mouse macrophages in vitro, Zbl., Bakt. Hyg., I.Abt. Orig. B 165, 260-268.
(19) Seemayer, N.H., N. Manojlovic and H. Behrendt (1980) Investigation on the cytotoxic effect of city smog extracts and their fractions on mouse macrophages in vitro, Europ. J. Cell Biol. 22, 1567.
(20) Sugiri, D., H. Behrendt and N.H. Seemayer (1985) Biological effects of city smog extract IX. Quantitative cytological investigation on the cytotoxic effect of city smog extracts on

macrophages in vitro, Zbl. Bakt. Hyg.I.Abt. Orig. B 181, 226-239.

(21) Behrendt, H., N.H. Seemayer and R. Tomingas (1986) Evidence for specific ultrastructural alterations of macrophages induced by airborne particulates. Aerosols: Formation and Reactivity. 2nd. Int. Aerosol Conf., Pergamon J. Ltd. Great Britain, pp. 212-214.

(22) Seemayer, N.H., W. Hadnagy, R. Tomingas and N. Manojlovic (1987) Cell- und genotoxic activities of airborne particulates from a highly industrialized region: A survey over a period of 11 years (1975-1986), J. Aerosol Sci. 18, 721-724.

(23) Seemayer, N.H., C.-C. Schürer, H. König and N. Manojlovic (1985) Mutagenicity of airborne particulate pollutants: Induction of point mutations and sister-chromatid exchanges in tissue culture cells, Mutation Res. 147, 134-135.

(24) Casto, B.C. (1980) Detection of chemical carcinogens and mutagens in hamster cells by enhancement of adenovirus transformation. in: Mammalian cell transformation by chemical carcinogens. Ed. by M. Mishra, V. Dunkel and M. Mehlman, Adv. in Modern Environ. Toxicol. Vol. I, Senate Press, Inc., Princeton Junction, pp. 241-272.

(25) Seemayer, N.H., N. Manojlovic and R. Tomingas (1986) Induction of malignant cell transformation in vitro by extract and fractions of airborne particulate matter. J. Aerosol Sci. 17, 356-360.

(26) Seemayer, N.H., R. Tomingas and N. Manojlovic (1986) Health hazards by airborne particulates: cytotoxic and genotoxic effects on mammalian cells in vitro. Aerosols: Formation and Reactivity. 2nd Int. Aerosol Conf., Pergamon J. Ltd, Great Britain, pp. 331-334.

2. Cytotoxicity and Genotoxicity on Bacterial Cells

THIAARENES - THEIR ENVIRONMENTAL OCCURRENCE AND BIOLOGICAL ACTIVITY

Jacob, J and G. Grimmer
Biochemical Institute for Environmental Carcinogens,
Lurup 4, D - 2070 Großhansdorf, FRG

INTRODUCTION

Polycyclic aromatic hydrocarbons (PAH) are a well known class of carcinogenic pollutants in the human environment. Their biological potential and metabolism have been thoroughly investigated. Thiaarenes, i.e. sulfur-containing polycyclic aromatic compounds (S-PAC) deriving from thiophene, often are associated with PAH especially in crude oil and in hard coal emission and exhibit very similar chemical properties. However, little efforts have been made to study the environmental relevance and the metabolism of thiaarenes (1). More recently, evidence for the carcinogenic and mutagenic activities of some thiaarenes has been presented (2, 3) and there are also strong indications that they may contribute to the carcinogenic potential of environmental matter, for instance in case of emissions from hard coal combustion (4).

MATERIALS AND METHODS

Enrichment of S-PAC

During the enrichment procedure thiaarenes remain in the PAH-fraction from which they may be separated by an oxidation/reduction process or by adsorption to palladium chloride.

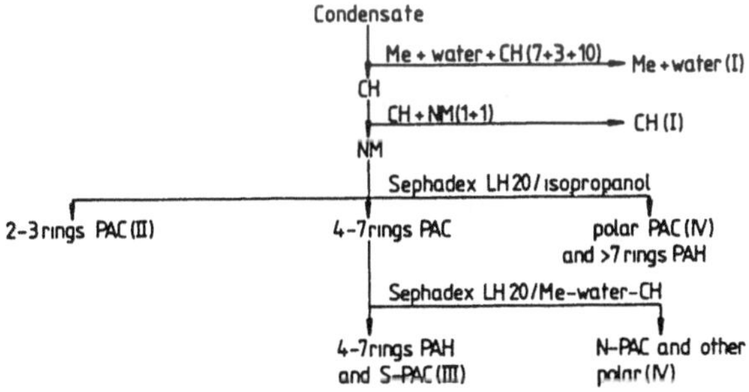

Figure 1: Scheme of PAH-enrichment from flue gas condendate

CH = cyclohexane
Me = methanol
NM = nitromethane
PAC = polycyclic aromatic compounds
N-PAC = nitrogen-containing PAC

Metabolic Studies

The preparation of rat liver microsomes and the incubation conditions have been published elsewhere (5).

RESULTS

The PAH-profile of the emission from hard-coal fired furnaces exhibits a total number of at least 119 different individuals, about one third of which could be characterized as S-PAC when recorded by a sulfur-selective GC-detector and/or analyzed by mass spectrometry.

Studying the carcinogenic activity of hard-coal emission, it became evident that this activity is almost entirely caused by the PAH-fraction (4) (see Figure 2).

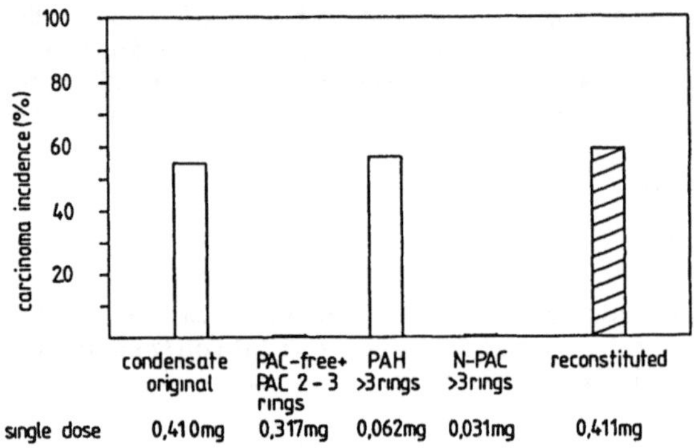

Figure 2: Balance of the carcinogenic activity of the emission from coal-fired furnaces. Dosage: 0.411 mg condensate twice a week. Fractions were dosed proportionately.

Further fractionation into a lower and a higher-boiling fraction evidenced that the carcinogenic potential is equally distributed to both fractions, i.e. that 4- and 5-ring systems on one hand, and 6 and more rings-containing systems on the other one, equally contribute to the carcinogenic activity of hard coal emission (4).

From various studies (3, 6) it is well known that the thiaarenes detected in environmental matter, are potent mutagens when activated by mammalian liver enzymes. We found that benzo(2,3)-phenanthro(4,5-bcd)-thiophene - an isoster to benzo(a)pyrene - exhibits an exceptional high mutagenic effet in Salmonella typhimurium TA98 und TA100 as well as in E. coli, being double as potent as benzo(a)pyrene itself (6). Although less efficient, other environmentally relevant thiaarenes were also found to be more or less potent mutagens.

Recently, Croisy et al. (7) tested some thiaarenes for carcinogenicity by subcutaneous injection into mice. A comparison to structurally related carbocyclic isosters shows that the thiaarenes in those cases investigated were more potent carcinogens than the PAH (Figure 3).

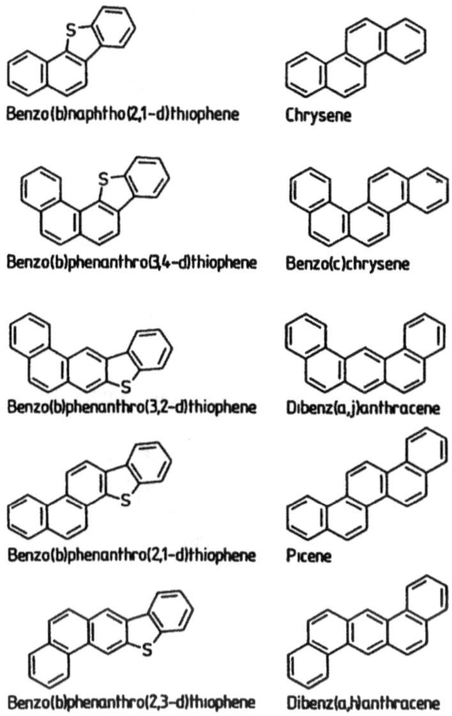

Figure 3: Some carcinogenic thiaarenes and their carbocyclic isosters.

Hence, it is to be expected that thiaarenes contribute markedly to the PAH-induced carcinogenicity of environmental matter.

Studies on the rat liver microsomal metabolism of thiaarenes were carried out to see whether the same pathways which were found for PAH are involved in the metabolic activation of S-PAC. However, in contrast to the epoxidation and subsequent hydrolysis of the epoxides to trans-dihydrodiols, which are the predominant metabolic activation steps of PAH-conversion with rat liver microsomes, thiaarenes are almost exclusively attacked at the sulfur atom, resulting in the formation of sulfoxides and sulfones. This has been demonstrated for 4 different compounds which all are widely distributed to the environment (Figure 4).

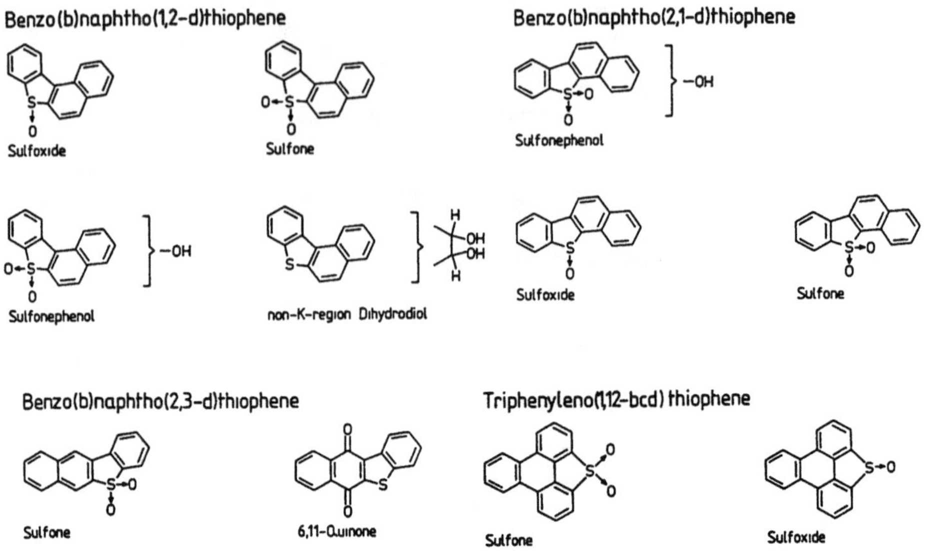

Figure 4: Rat liver microsomal metabolites of benzo(b)naphtho(1,2-d)-, -(2,1-d)-, (2,3-d)thiophene, and triphenyleno(1,12-bcd)thiophene.

Ring oxidation plays only a minor role in all of the 4 substrates if at all. However, when the various main metabolites of these thiaarenes - all 4 sulfones and sulfoxides and the 6,11-quinone of the benzo(b)naphtho(2,3-d)thiophene - were tested for mutagenicity, it was found that the sulfone of benzo(b)naphtho(2,1-d)thiophene was a potent direct acting mutagen and hence, possibly the biological active form of this thiaarene (100 rev./1 µg).

Although very preliminary, the results show that thiaarenes might be considered as possibly hazardous compounds to humans.

SUMMARY AND CONCLUSIONS

Thiaarenes are associated with PAH in environmental matter. Some of them are potent mutagens and carcinogens. In contrast to PAH, in thiaarenes rat liver microsomal oxidation takes place at the sulfur atom rather than at the carbocyclic part of the molecule resulting in the formation of sulfoxides and sulfones. Thiaarenes are expected to contribute to the carcinogenic activity of environmental matter.

REFERENCES

(1) Jacob, J., A. Schmoldt and G. Grimmer (1986) The predominant role of S-oxidation in rat liver metabolism of thiaarenes, Cancer Lett., 32, 107-116.

(2) Waravdekar, S.S. and K.J. Ranadive (1957) Biological testing of sulfur isosters of carcinogenic hydrocarbons, J. Natl. Cancer Inst., 18, 555-563.

(3) Pelroy, R.A., D.L. Stewart, Y. Tominaga, M. Iwao, R.N. Castle and M.L. Lee (1983) Microbial mutagenicity of 3- and 4-ring polycyclic aromatic sulfur heterocycles. Mutat. Res., 117, 31-40.

(4) Grimmer, G., H. Brune, R. Deutsch-Wenzel, G. Dettbarn and J. Misfeld (1987) Contribution of polycyclic aromatic hydrocarbons and polar polycyclic aromatic compounds to the carcinogenic impact of flue gas condensate from coal-fired residential furnaces evaluated by implantation into the rat lung, J. Natl. Cancer Inst., 78, 935-942.

(5) Jacob, J., A. Schmoldt and G. Grimmer (1982) Influence of monooxygenase inducers on the metabolic profile of phenanthrene in rat liver microsomes, Toxicology, 25, 333-343.

(6) Karcher, W., A. Nelen, R. Depaus, J. van Eijk, P. Glaude and J. Jacob (1981) New results in the detection, identification and mutagenic testing of heterocyclic polycyclic aromatic hydrocarbons, in: Polynuclear Aromatic Hydrocarbons (M. Cooke and A.J. Dennis, eds.), pp. 317-327, Battelle Press, Columbus/Ohio.

(7) Croisy, A., J. Mispelter, J.M. Lhoste, F. Zajdela and P. Jacquignon (1984) Thiophene analogues of carcinogenic polycyclic hydrocarbons. Elbs pyrolysis of various aroylmethylbenzo(b)thiophenes. J. Heterocycl. Chem., 21, 353-359.

HEAVY METALS AND POLYAROMATIC COMPOUNDS IN SUSPENDED PARTICULATES OF DIFFERENT EMISSION SOURCES AND THEIR MUTAGENICITY

Moriske H.-J.and H. Rüden
Institute for Hygiene at the Free University of Berlin
Hindenburgdamm 27, D-1000 Berlin 45, F.R.G.

INTRODUCTION

In the past, some organic and inorganic compounds in urban suspended particulates at different measurement stations in Berlin(West) were determinated (2,4,5).
Of the organic fraction, a mutagenicity test was done using the Salmonella typhimurium bacteriological bioassay by AMES et al. (1,3).
In these immission samples, some differences between the measurements were shown in dependence of the location of the measurement station and in dependence of the influence of different emission sources which were mainly automobile traffic and domestic fuel sources (4,5).
Therefore, in further investigations suspended particulates should be sampled directly from the emission sources.

METHODS

During April and May 1985, we measured the emissions from a stove-heating system which was fired with bituminous and anthracite coal. The emission was measured nearby the chimney output at different conditions of the combustion process. For the measurements, high volume samplers with csacade impactors were used which allowed a fractionation of the suspended particulates into particles sizes from 10.2-2.1 µm < 2.1 µm.
To avoid condensation processes and high gas temperatures a dilution with surrounding air was done (about ten times).

Between June and November 1985, the emission of suspended particulates from vehicle engines was measured using a special tunnel dilution equipment at different engine conditions. Here, also a fractionation into different particle sizes was done.
At last, during September 1985, suspended particulates were collected in a highway traffic tunnel (length of about 2 km).

For the identification of the organic compounds, an extraction of the filter material with diethylether/cylcohexane was done.
From the etherextractable organic matter (= EEOM), we made a separation into neutral, basic and acidic compounds using several basic and acidic agents as reagents.
The neutral fraction was separated by thin layer chromatography further into aliphatic compounds, polycyclic aromatic hydrocarbons (= PAH) and polar neutral organic compounds (= POCn) which include mainly oxygenated PAH, like ketones, nitro-PAH and others.
The concentration of single organic compounds from the PAH and POCn-fraction was determined by gas chromatographic analysis using fused silica capillary columns (SE-52, OV-1, FFAP) and FID-detection. The determination was done using internal standard methods.

Some filters were extracted in another procedure for making an identification of heavy metal elements.
The filter material was cooked with nitric acid (HNO_3) (35%). The solution was filtered and dried and then resolved in HNO_3 (1%).
The concentration of the heavy metal elements was determined using atom absorption spectrometric method and flameless analysis. 9 heavy metal elements were identified.

All organic subfractions and the whole etherextractable organic extract were tested in the Salmonella mammalian microsome baceriological bioassay by AMES et al. (1) and MARON et al. (3), using the bacterial strains TA98, TA100 and TA98NR. The dose concentrations were calculated after dose-level tests and ranged from 100-200 µg/plate for each fraction.

RESULTS AND DISCUSSION

Investigating the amount of the total suspended particulates (TSP) in relation to different particle sizes, we found the main part (in %) of the TSP for particle sizes < 2 µm aerodynamic diameter (up to 90% in emission samples and 60% in the highway tunnel).

For the etherextractable organic matter (EEOM) and organic subfractions, we found that the main part of the organic matter was extracted in particle sizes < 2.1 µm aerodynamic diameter.
In the emission samples, about half amount of the TSP (40-55%) were extractable organic compounds whereas in the highway tunnel we got lower amounts of the EEOM (16%). Separating the EEOM, the most compounds were extracted in the neutral fraction (more than 75%); in the acidic fraction we found 15-19% and in the basic fraction we found only 6-13%.
The neutral fraction was separated into aliphatic compounds, PAH and POCn. From these table 1 shows the results of the chromatographic analysis of PAH and POCn compounds (average values from single measurements) as percentages of the total amount of PAH and POCn which showed different percentages relating to the whole amount of identified compounds between the two emission sources.

Table 1: Percentages of PAH and POCn in relation to the whole amount of all identified PAH and POCn (= 100% each)

PAH/POCn	domestic fuel	diesel engine	highway tunnel
acenaphthene	1.6	0.1	0.2
phenanthrene	18.9	2.6	3.1
perylene	2.6	4.8	7.3
benzo(g.h.i)perylene	2.0	9.6	6.6
coronene	1.6	4.0	5.6
2-nitronaphthalene	4.4	0.1	0.3
9-fluorenone	17.2	6.7	7.9
acenapthenchinone	1.7	16.4	17.0
anthrachinone	2.3	21.3	5.9
1-nitropyrene	3.5	18.4	15.6
2.7-dinitrofluorene	3.3	0.8	0.0

The main identified compounds were in both cases fluoranthene (16.4-32.9%), pyrene (3.8-15.8%), and for diesel emissions also chrysene (6.6-15.6%). Differences between the sources were shown mainly for acenaphthene (0.1-1.6%) and phenanthrene (2.6-18.9%) which showed higher percentages in coal-firing and for perylene (2.6-7.3%), benzo(g.h.i)perylene (2.0-9.6%) and coronene (1.6-5.6%) which showed higher percentages in diesel emissions and highway tunnel samples.
For the POCn, we got higher percentages in coal firing mainly for 9-fluorenone (6.7-17.2%) and 2-nitronaphthalene (0.1-4.4%). In contrary, higher percentages in vehicle emissions and measurements in the highway tunnel we got for acenaphthenchinone (1.7-17.0%), anthrachinone (2.3-21.3%) and mainly 1-nitropyrene (3.5-18.4%). The amount of the more nitrated compounds was often higher in the coal firing emissions.

Also for the concentrations of the heavy metals we found some differences between the two emission sources. Table 2 shows the results.
The concentration of 8 heavy metals was standardized to the concentration of iron. The value of iron was fixed to 10.00.

Table 2: Concentration index of heavy metals in relation to the concentration of iron (=10.00)

heavy metal element	domestic fuel	diesel engine	highway tunnel
iron	10.00	10.00	10.00
manganese	0.04	0.27	0.24
lead	3.64	4.04	5.37
cadmium	0.03	0.09	0.01
nickel	0.08	0.25	0.07
chromium	0.27	0.38	0.05
vanadium	0.26	0.23	0.02
arsenic	1.40	0.02	0.12
selenium	0.30	0.04	0.07

The indexes of the elements show that we got higher amounts for manganese, lead and partially nickel and chromium in vehicle emissions whereas in domestic fuel emissions we often found higher amounts for lead, chromium, vanadium and, in comparison with the vehicle emissions, especially higher amounts for arsenic and selenium.
The high concentration of arsenic was found mainly when we used anthracite coal for firing.
The results of the mutagenicity tests are shown in table 3 for the strain TA 98. The highest mutagenic activity was found for the acidic, PAH- and POCn-fraction, generally. The basic fraction showed a higer mutagenic activity only in domestic fuel emissions.

Table 3: Mutagenic activity of organic subfractions (revertants/µg); strain TA 98; f. = fraction,+/- S9 = with/without metabolic activation

particle size (µm)	>10.2		10.2-2.1		>10.2		10.2-2.1	
metabolism	+S9	-S9	+S9	-S9	+S9	-S9	+S9	-S9
EEOM	0.90	0.62	1.27	0.94	0.29	0.19	0.52	0.44
neutral f.	0.91	0.72	1.26	0.73	0.32	0.16	0.94	0.61
acidic f.	1.00	0.99	1.22	0.95	0.52	0.34	1.78	1.52
basic f.	0.83	0.66	1.20	0.66	0.31	0.16	0.60	0.30
aliphatics	0.21	0.21	0.26	0.15	0.24	0.18	0.20	0.17
PAH	1.10	0.71	0.89	0.73	0.61	0.33	0.66	0.51
POCn	1.78	0.97	1.63	1.07	1.35	0.51	2.51	1.18

SUMMARY AND CONCLUSIONS

In summary of all the results of the organic and inorganic analysis, we found that both emission sources, vehicle diesel engines and domestic fuel emissions which means here coal-firing, are responsible for the emisson of a lot of toxic and mutagenic suspect agent in urban air.
The emission profiles of single substances often were similar but also for some elements and organic compounds were quite different. This perhaps gives a possibility to distinguish the influence of both emission sources also in the immission aerosol.
The AMES-bioassay can be used not only for detect mutagenic suspect agents but also for distinguish emission influences without knowing the exact composition of the organic extract by using selective bacterial strains like nitroreductase-deficient strains or others. For getting more detailed informations about the general health risk, tests in mammals and epidemiological studies are necessary.

AKNOWLEDGEMENTS

The authors appreciate the financial sponsorship of the Deutsche Forschungsgemeinschaft (DFG).

REFERENCES

(1) Ames, B., J. Mc Cann and E. Yamasaki: Methods for detecting carcinogens and mutagens with the Salmonella mammalian microsome mutagenicity test. Mutation Res. 31 (1975) 347-364

(2) Lieback, J.U.: Staub- und Schwermetallkonzentrationen im Berliner Stadtaersosol und deren Bewertung aus der Sicht der Lufthygiene. VDI-Fortschrittberichte, Reihe 15 Nr.34 (1985) 1-277

(3) Maron, D.M. and B.N. Ames: Revised methods for the Salmonella mutagenicity test. Mutation Res. 113 (1983) 173-215

(4) Moriske, H.-J.: Polare organische Verbindungen im Stadtaerosol: Chemische Zusammensetzung und mutagene Wirksamkeit im hausbrand- und Kfz.-belasteten Aerosol. VDI-Fortschrittberichte, Reihe 15 Nr.42 (1986) 1-354

(5) Moriske, H.-J., I. Block, H. Schleibinger and H. Rüden: Polar neutral organic compounds (POCn) in urban aerosols 1. communication: Chemical characterization and mutagenic activity influenced by different sources. Zbl.Bakt.Hyg., I.Abt.Orig.B 181 (1985)240-271

MICROBIOCIDAL EFFECTS OF EXHAUST GASES AND THEIR SECONDARY PRODUCTS UNDER FIELD CONDITIONS

Waldner-Sander, S., Botzenhart, K.
Hygiene-Institut der Universität Tübingen, Abt. für Allgemeine Hygiene und Umwelthygiene, Silcherstr. 7, D-7400 Tübingen, FRG

INTRODUCTION

This paper presents one part of our investigations, in which microorganisms are used as bioindicators for airborne pollutants. The feasibility of employment of various microorganisms as indicators has been demonstrated in former papers. The deleterious influence of ozone on the viability of microorganisms and the existence of toxic compounds in outdoor air in densely populated and also in remote areas could be evaluated (1). Furthermore toxic and mutagenic effects of exhaust gases could be shown. Such effects are also often described in literature, obtained from various laboratory experiments by means of manifold test systems. Compared with this we could record the influence of the sum of the constituents on living organisms with a method, which was developed to quantitate the toxic effects of air pollutants using specially selected microorganisms in open air experiments. Contrary to other bioindicator systems only a very short exposure time (90 min) is needed.

METHODS AND SAMPLING SITES

Prior to exposure of the test strains to the open air at a given site, two sampling chambers were prepared as follows: Membrane filters were loaded with about 100 cells of a given test strain. 30 filters of each strain were placed in the incubation chambers. Subsequently, the chambers were adjusted to the desired relative humidity, sealed, purged with synthetic air and transported to the exposure site. After a minimum exposure time of 30 min, transfer of the membrane filters of one chamber to agar plates was initiated and continued at regular intervals up to a maximum exposure time of 90 min. Immediately thereafter, filters which served as controls were removed from the second chamber which remained sealed up to this time. After 1 or 2 days incubation, colonies were counted and the reduction rate λ was calculated from the difference between control and exposure experiments. The test strains were exposed at 3 measuring sites in the Schönbuch forest near Bebenhausen. The first site was situated near a high frequented road beside a monitoring station, which is equipped with instrumentation for monitoring wind velocity and -direction and levels of sulfur dioxide, nitrous oxides, ozone and carbon monoxide. The second site was located at the edge of the forest in a distance to the street of about 500 m. The third site was within the forest area in a distance of 1000 m.
The typical immission situation at the monitoring station on sunny days without wind shows increasing levels of ozone to over 100 $\mu g/m^3$ during the day. The highest values are recorded between 12 and 16 o'clock. Afterwards they go back below a level of 30 $\mu g/m^3$. The levels of the compounds NO_x and CO reflect here increased motor vehicle pollution at hours of increased traffic density. They are very low during the afternoon. Formation of NO_2 from NO correlates with the level of ozone. Parameters like wind velocity and wind

direction could have great influence on the concentration because of dilution effects.

RESULTS

In figure 1 the reduction rates of three testorganisms on a few days in summer 1986 are demonstrated. It was surprising, that the greatest loss of viability was not seen in direct vicinity to the street, but in some distance in the forest area. Essentially, weather conditions were responsable for the increase. On rainy days all organisms showed low death rates. (e.g. on August 28 th, only 0,5 hours of sunshine and rain fall of 4,5 liter /m² were recorded). A low reduction rate was also obtained, when it was raining some hours before exposure. Sunny days revealed clearly increased values at the edge of the forest. On august 14 th and on august 20th there were more than 7,3 hours of sunshine. On both days no rain was observed.
High differences insensitivity of various microorganisms to noxious agents is also demonstrated in this diagrams. This could also be documented in earlier papers (2). The yeast Candida albicans tolerated outdoor exposition better than the other two bacteria species.

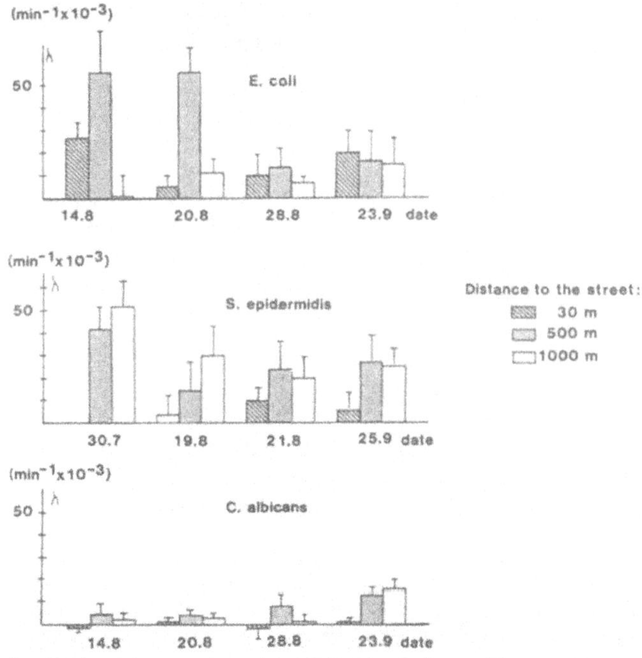

Fig. 1: Reduction rates λ, obtained from the exposure of Escherichia coli, Staphylococcus epidermidis and Candida albicans in summer 1986 at the measuring sites in Bebenhausen.

In autumn, the results vary considerably (fig. 2). In spite of 7,4 hours of sunshine on october 27 th and still 5,9 hours on november 13 th and without rain on these days and the day before, only low reduction rates could be observed, for both E. coli and Candida albicans. These are common results in the cold season. Remarkable values were obtained only in summer. There were also never any significantly increased death rates in the forest area in autumn.

The following hypothesis tries to explain these results: Some compounds released by combustion processes are highly toxic to micro-organisms. They are very reactive and in a distance of a few meters there is mainly aged air in respect of this anthropogenic components. The taint of anthropogenic compounds in forest air is rather high. Stable substances which exhibit long lifetimes under natural conditions are dominating. They include a number of hydrocarbons and various alkylbenzenes both of which are released by combustion processes. Other substances also detected were esters and halogenated compounds which are widely used as solvents. The dense population in the region and its manifold emission sources account for the transport of polluted air into the forest areas.

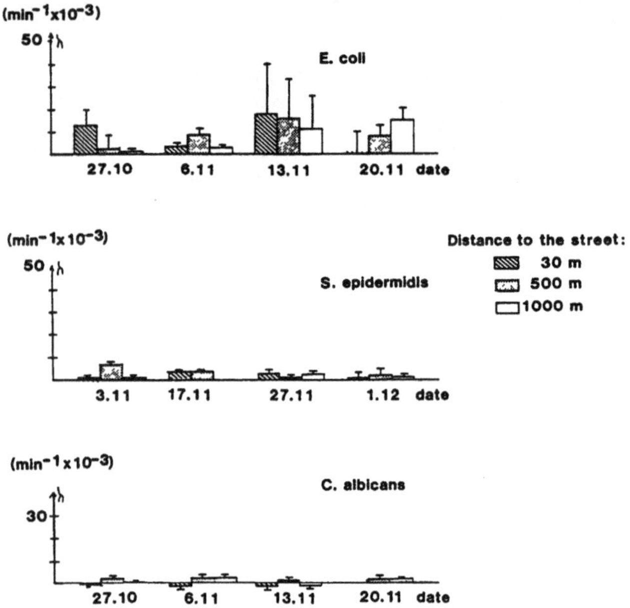

Fig. 2: Reductions rates λ received by exposing Escherichia coli, Staphylococcus epidermidis and Candida albicans at the measuring sites in Bebenhausen in autumn 1986.

In the forest area, however, other compounds from natural sources are found which are photosmôg precursors because of their reactivity against anthropogenic pollution products. These biogenic components, for example isoprene and various monoterpenes, could be measured in amounts in which they can contribute essentially to the dynamics of air chemistry (3).
The occurence of the biogenic compounds in Bebenhausen compared with the microbial reduction rates showed a very good correlation. Biogenic compounds reach many times higher concentrations in summer than in the cold season. The liberation of the compounds is influenced by season and also daily variations to a high degree. A further, very interesting point is the rapid decrease of concentrations of biogenic olefines in the distance to the forest. This may explain the higher values of the death rates in the forest area in summer.

To underline the toxic effect especially of photooxidants, the reduction rates at different hours of the day are demonstrated in figure 3. The highest values were observed at times of the optimal conditions of photochemical reactions. The reduction ratesare not influenced by temperature. This could be excluded in laboratory experiments. There is no effect of temperature on all organisms in the short exposure time of 90 min. Furthermore not only biogenic compounds are responsable for the higher reduction rates in the forest area. In areas of higher elevation and therefore a more intensive radiation an increased loss of viability could not be observed. The reduction rates at a measuring site in an elevated, rural area in a clean air region (900 m) are lower in the annual mean than at the site in Bebenhausen (380 m).

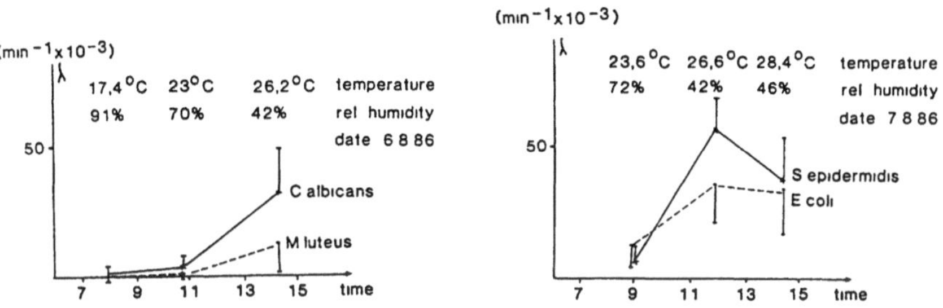

Fig. 3: Reduction rates λ, obtained by exposure of the test organisms during 2 days in Bebenhausen.

SUMMARY AND CONCLUSIONS

The present results demonstrate the feasibility of the employment of microorganisms as indicators of airborne pollutants, especially photooxidants, and the advantage of using different test strains simultaneously to monitor a wide range of potential toxic effects of air pollution.

We found increased death rates at all sites affected by motor vehicle exhaust gases and urban air pollution in contrast to rural locations. In forest areas near densely populated regions a loss of viability was established on days when photooxidants and biogenic olefines could be expected.

Because of the demonstrated rapidity with which the presented microbial bioindicators,provide information about air quality, it is possible to draw conclusions about the dynamics of apperance of toxic air pollutants.

REFERENCES

1. Waldner-Sander,S., Botzenhart, K.: Mikroorganismen als biologische Indikatoren für waldschädigende Luftverunreinigungen. KfK-PEF Nr 4, 225-227 (1986).
2. Jüttner. F.: Quantitative analysis of monoterpenes and volatile organic pollution products (VOC) in forest air of the southern Black Forest. Chemosphere (in press).
3. Waldner-Sander, S., Botzenhart, K.: Microorganisms as biological indicators of air pollution. In: Advances in Aerobiology, EXS 51, 395-400, Birkhäuser Verlag Basel (1987).

3. Specific Effects on Cell Functions

Tumor Promoter PMA and Crosslinking of surface Immunoglobulins induce Superoxide production in Epstein-Barr-Virus-infected Human B Lymphocytes : a novel pathway for the production of potentially mutagenic species

Maly,F.-E.,A.R. Cross,O.T.G. Jones and A.L. de Weck
Institute of Clinical Immunology,Inselspital Bern,-Schweiz and Department of Biochemistry,University of Bristol,United Kingdom

INTRODUCTION

Superoxide anion ($^{-}O_2$) and derived reactive oxygen species like hydroxyl radical are important potentially mutagenic species.In biological system,they can be produced by phagocytes triggered by contact with particles,but also under the influence by different soluble mediators (1).This includes the complement component C5a and certain cloned cytokines like tumor necrosis factor and lymphotoxin (2); in addition,supernatants of mononuclear cells treated with either Lipopolysaccharide(3)or with Silica particles (4) contain granulocyte-activating factors probably different from those already mentioned.
However,Epstein-Barr-Virus-transformed B lymphocyte lines (EBV-BLCL)have also recently been recognized as producers of superoxide,giving the first clear indication that lymphoid cells may also represent a source of reactive oxygen species(5).We have compared the superoxide generating system of such EBV-BLCL with the phagocytic NAD(P)H-oxidase by assaying EBV-BLCL for the presence of cytochrome b_{-245} and a 45kDa Diphenylene-binding peptide (DPI-binding peptide),both specific components of NAD(P)H-oxidase(6).
Further,in the search for a physiological counterpart of EBV-BLCL,we have examined tonsillar cell preparations for the presence of natural,non-transformed B lymphocytes capable of superoxide production.

MATERIALS AND METHODS

EBV-BLCL :The methods used for the characterisation of the superoxide generating chain of EBV-BLCL have been reported in detail (6).Tonsillar cells were obtained from surgical tonsillectomy specimens (courtesy of Prof.M.Neiger,Department of E.N.T.,Inselspital Bern) by gentle mechanical dispersion to a single-cell suspension.Estimation of the frequency of superoxide-producing cells was done by reduction of nitroblue-tetrazolium .Cell surface marker analysis was performed on a model 50H Cytofluorograf (ORTHO) with

FITC-labelled monoclonal antibodys specific for T lymphocytes (anti-CD3),B lymphocytes (anti-B1),Monocytes/Macrophages (anti-M3) and Granulocytes.

RESULTS

The properties of the superoxide generating system of EBV-BLCL ("B cell oxidase" are compared with those of the NAD(P)H-oxidase of granulocytes (PMN) in table 1.Evidently,the "biochemical machineries" responsible for superoxide production appear similar,but the modes of activation are quite distinct.While a comparatively large number of stimuli trigger an oxidative burst in PMN,only PMA and the crosslinking of surface Immunoglobulins are able to activate the "B cell oxidase".

Table 1 Properties of "B cell oxidase" in comparison with NAD(P)H-oxidase of neutrophilic granulocytes [1)]

	Neutrophils	EBV-BLCL
Presence of Cytochrome b_{-245}	+	+
Presence of DPI-binding peptide	+	+
Inhibition of $^{-}O_2$ - production by DPI	+	+
Stimulation of $^{-}O_2$ - production by :		
Zymosan particles	+	0
C5a	+	0
Tumor necrosis factor	+	0
PMA	+	+
Protein A-Staphylococci	N.D.	+

+ indicates positive response,0 no stimulation observed. [1)] Published in detail in (6).

Tonsillar Leukocytes typically contain predominantly B lymphocytes,while only low concentrations of Monocytes/Macrophages and Granulocytes are found.However after stimulation with PMA or Protein A-staphylococci,the frequency of NBT-positive superoxide-producing cells is well in excess of the combined frequencies of phagocytic cells (Table 2).

Table 2 . Cellular composition and frequency of superoxide-producing cells in tonsillar leukocytes.

T - Lymphocytes	33 %
B - Lymphocytes	73 %
Monocytes/Macrophages	0.5 %
Granulocytes	0.4 %

NBT-positive cells after stimulation with :

Control (Medium)	0.3 %
PMA (20 ng/ml)	15.1 %
Protein A - Staphylococci ("Pansorbin",Calbiochem)	14.2 %
Control Staphylococci ("Sansorbin",Calbiochem)	1.1 %

Staphylococci were used at 1mg/ml.Data shown are from a typical experiment.NBT reduction was completely suppressed by Superoxide dismutase (100 µg/ml).

DISCUSSION

The superoxide producing system of EBV-BLCL is structurally clearly very similar,if not identical to the "phagocytic" NAD(P)H-oxidase.However,its mode of activation appears to be different,since only PMA and the crosslinking of surface immunoglobulins have so far been shown to be effective stimuli.
Since surface immunoglobulins serve as antigen receptors in B cells,it is possible that under appropriate conditions,antigen may also be able to trigger superoxide production in EBV-BLCL.
Clearly,B-cell rich tonsillar leukocyte preparations contain sueroxide producing cells well in excess of the numbers of phagocytic cells.While we have not explicitly shown that these cells are in fact B cells and not T cells,this is suggested by the equal efficieny of PMA and Protein A-bearing staphylococci in inducing NBT reduction.It appears therefore ,that a subset of normal,non-transformed B lymphocytes possesses the capacity to produce superoxide in a fashion analogous to EBV-BLCL.

Superoxide production by such "superoxide-producer B cells" might be of importance in inflammation,but might also exert mutagenic effects.Such effects might be at play in the known phenomenon of somatic hypermutation in B cells,which is a key process in the generation of antibody diversity and affinity maturation (7).However,superoxide production by B cells might also be related to the known higher frequency of neoplastic transformation of B cells to lymphomas as opposed to T lymphocytes.

REFERENCES

(1) Babior,B.M. (1978).Oxygen-dependent killing by phagocytes.N.Engl.J.Med. 298:659-668 and 721-725.

(2)Klebanoff,S.J. et al.(1986).Stimulation of neutrophils by tumor necrosis factor.J.Immunol. 136 :4220-4225.

(3) Maly,F.-E. et al.(1986).Induction of Granulocyte chemiluminescence by a mediator derived from human monocytes.Lymphokine Res.5(1):21-33.

(4)Maly,E.R. et al. (1987) , cf this volume

(5) Volkman,D.J. et al. B cell lines as models for inherited phagocytic disease.J.Immunol.133:3006-3009.

(6) Maly,F.-E. et al. (1987)Biochemical and functional properties of the superoxide generating system of B cell lines.Submitted,J.Immunol.

(7) Griffiths,G.M et al. (1984) Somatic mutation and the maturation of immune response to 2-phenyl-oxazolone.Nature 312:271-275.

STRONG EFFECTS OF CYTOKINES FROM QUARTZ DUST EXPOSED HUMAN MONONUCLEAR CELLS ON HUMAN MESENCHYMAL CELLS AND POLYMORPHONUCLEAR LEUKOCYTES

Maly, E., N. H. Seemayer, Heidrun Behrendt, N. Manojlovic , Agnes Braumann and F. E. Maly
Medical Institute of Enviromental Hygiene at the University of Düsseldorf, Gurlittstr. 53,D-4000 Düsseldorf, FRG
and * Institute for Clinical Immunology, Inselspital Bern, CH-3010 Bern, Switzerland

INTRODUCTION

Different cytokines are involved in the immune response in a very manifold manner as became evident in the last few years. We report about cytokines from quartz dust exposed human mononuclear cells which may be important in the development of silicosis. In chronic experimental silicosis, pulmonary fibrosis is associated with an intense inflammatory response. The predominant cell in the interstitium and the airspaces is the macrophage. But neutrofils also accumulate in the silicotic lung, especially in the early phases. Mechanisms, by which inflammation is initiated and modulated in the lung exposed to silica are poorly understood. Therefore, we investigated the hypothesis, that nondirect cascade mechanisms of quartz toxicity may be involved in human lung tissue damage. Several reports indicate that monocytes / macrophages produce and release a number of mediators after contact with quartz dust in both human and animal systems (3). These mediators include interleukin - 1, which is a potent inflammatory cytokine (6). Granulocytes, appearing early in rat alveoli after inhalation of quartz dust DQ-12 (2), might contribute to quartz induced tissue damage . However, quartz dust DQ-12 has been reported not to be stimulatory for granulocytes by itself. Conceivably, contact of macrophages with quartz dust DQ-12 could lead to the liberation of mediators, which could then either activate the toxic potential of granulocytes and/or stimulate the growth of fibroblast cultures. Since granulocyte toxicity is largely dependent on their generation of free oxygen-radicals, we investigated whether supernatants of quartz dust DQ-12 treated mononuclear cells could stimulate granulocyte free radical production.

MATERIALS AND METHODS

Human monocytes/macrophages and human granulocytes were isolated from the peripheral blood by density gradient centrifugation as described by us earlier (4). For supernatant production 4 x 10^6 /ml monocytes/ macrophages were incubated with different concentrations of quartz-dust DQ-12 (Dörentruper Kristallquarzmehl, Mahlung No. 12, grain size < 5 um). After 24 hours supernatants were collected, centrifugated and filtrated through GV-Millipore-membranes.
Granulocyte free radical production was measured as lucigenin dependent chemiluminescence in a 6-channel-luminometer (LB 9505, BERTHOLD, Wildbad) or in a HAMAMATSU C-1966 Video-System (fig.1).
Fibroblast stimulating activity was determined on human FH-3 fibroblasts cultured in Dulbecco's MEM with 50 % of mononuclear cell supernatant. After 6 days of cultivation, protein content of control and stimulated cultures was measured by a modified Lowry method.

RESULTS

Supernatants of quartz dust DQ-12 treated mononuclear cells were able to stimulate granulocyte chemiluminescence. Granulocyte stimulating activity was however strictly dependent on the dose of quartz dust DQ-12 used : This effect was demonstrable from 50 - 200 ug/ml DQ-12 . Strongest granulocyte stimulating activity was found in supernatants produced by 50 ug/ml DQ-12 (fig.1). Superoxide dismutase at a final concentration of 1 mg/ml inhibited more than 75 % of the response, suggesting that superoxide production was responsible for the observed chemiluminescence (fig.2). When supernatants were prepared at optimal conditions (50 ug/ml DQ-12), the granulocyte stimulating activity was demonstrable up to the dilution 1:16 (fig.3). Activity of supernatants could be progressivly diminished by heat treatment and was abolished after boiling supernatants for 60 minutes To approach the question whether granulocyte stimulating activity of the supernatant was due to (a) mediator(s) of protein character, active supernatant was treated with various enzymes. For this investigation immobilized enzymes were used because addition of such enzymes to granulocytes might have an influence on granulocyte activity. Loss of the granulocyte stimulating activity of supernatants was observed only with bacterial protease (81 % loss) and chymotrypsin (55 % loss), whereas neither treatment with trypsin, ribonuclease or neuraminidase caused significant changes. When cycloheximide (20 ug/ml) ,an inhibitor of protein synthesis, was added to the cultured mononuclear cells, granulocyte stimulating activity was completely deminished. Therefore it would appear that the mediator has protein character. We observed fibroblast stimulating activity over a broad dose range of quartz dust DQ-12 used for supernatant production. Fibroblast stimulating activity could be demonstrated from 30 -240 ug/ml DQ-12 Activation of granulocytes by the supernatant could also be shown when using the NBT-test (Nitrobluetetrazolium-reduction capacity). When studying the activated granulocytes with an electron microscope, we found strong chemotactic changes in their morphological structure. In contrast to the control cells granulocytes which had been incubated with the supernatant of quartz-dust DQ-12 treated human monocytes/macrophages for 15 min. show strong chemotactic activity as can be seen by changes of the cell shape and development of a leading lamella. After an incubation period of 60 min. chemotactic activity is still present, but reduced in comparison to 15 min. of incubation. Additionally the cell seems to swell to some extend and there is a reduction of specific intracytoplasmatic granules, indicating the release of lysosomal products. Therefore, alterations of granulocytes induced by supernatants of quartz-dust DQ-12 treated monocytes/macrophages can be demonstrated by ultrastructural analysis in a time dependent manner.

DISCUSSION

The results shown above present an in vitro example of a possible nondirect mechanism of quartz dust toxicity. Mononuclear cells, when treated with low amounts of quartz dust DQ-12, elaborate a soluble cytokine-like mediator, which then in turn can activate granulocytes for increased production of reactive oxygen species. Oxygen species could either be directly toxic and thus lead to alveolar damage, or could function to increase activity of prolylhydroxylase, an important enzyme in collagene synthesis (1). The chemotactic effects of the

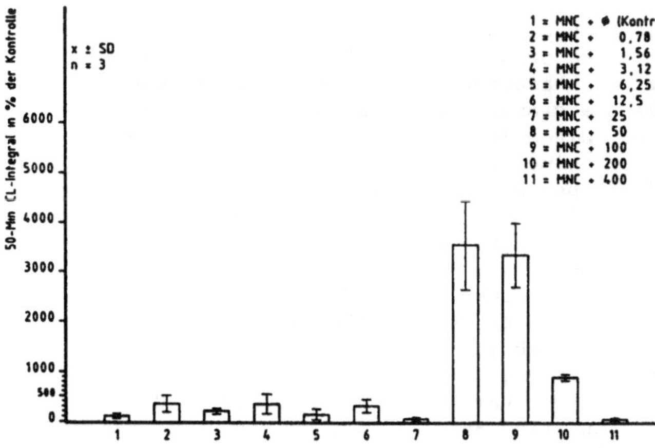

<u>Fig. 1 :</u> Effect of quartz dust DQ-12 concentration of exposed human mononuclear cells on the "granulocyte-stimulating activity" of supernatants

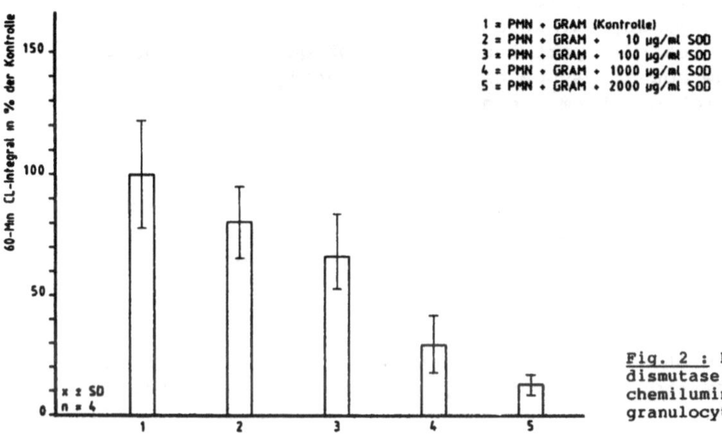

<u>Fig. 2 :</u> Effect of Superoxide-dismutase on GRAM-induced chemiluminescence of human granulocytes

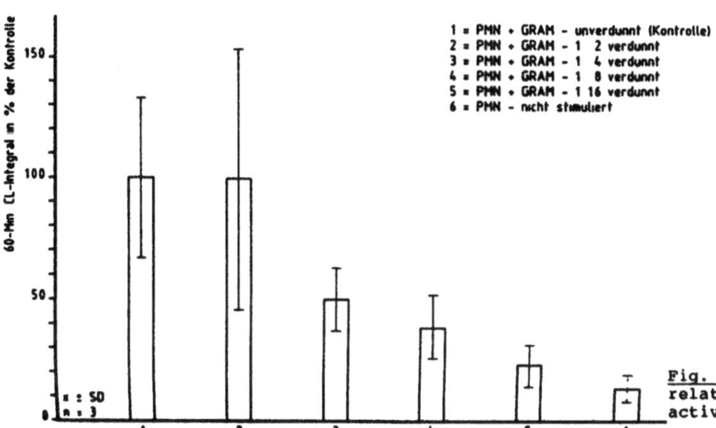

<u>Fig. 3 :</u> Dose-effect-relationship of granulocyte-activating mediators (GRAM)

supernatants on granulocytes may declare the appearence of granulocytes in the early phases of silicosis. The exact cellular origin and further biochemical characteristics of this quartz dust DQ-12 induced mediator having a granulocyte stimulating activity (GSA) must be elucidated in future. Since it is known that monocytes can produce interleukin - 1 after stimulation with silica, it is necessary to relate the phenomenon described by us to interleukin - 1 (m.w. 17-19 kda). Interestingly, studies with recombinant human interleukin-1 have shown, that interleukin - 1 (alpha or beta) does not trigger an oxidative burst in human granulocytes (5).Consequently, it is unlikely that interleukin - 1 is responsible for quartz induced granulocyte stimulating activity which we describe. In contrast, it is possible that the observed fibroblast stimulating activity of quartz induced supernatants is at least in part due to interleukin-1. Therefore, further efforts towards characterisation of quartz dust DQ-12 induced granulocyte stimulating activity are needed. Also the relevance of our in vitro findings for the in vivo situation will have to be addressed. Recently it was observed that alveolar macrophages react in a similar way to quartz dust DQ-12 as we have described for blood monocytes. In summary, our data support the hypothesis that at least in part the toxicity of quartz dust DQ-12, particularly at low doses, may be caused nondirectly by induction of inflammatory cytokines from mononuclear cells.

REFERENCES

1. Bathnagar, R. S.: The role of superoxide in oxidant-induced pulmonary fibrosis. Biochemical effects of Environmental Pollutants Ed. by S. D. Lee. Ann Arbor Science Publ., Ann Arbor, Mich., 1977, p.47/58

2. Behrendt, H. und K. H. Friedrichs : Effect of low dose and short-time inhalation of silica dust on the morphology of rat lungs. Aerosols : Formation and Reactivity, Pergamon Journals Ltd., GB, 1986, S. 279/282

3. Davis, G. S. : Pathogenesis of Silicosis : Current Concepts and Hypotheses. Lung, Volume 164, Nummer 3, 1986, S. 139-154

4. Maly, E., N. H. Seemayer, Heidrun Behrendt, N. Manoljlovic , Agnes Braumann and F. E. Maly : The importance of immunmodulative cytokines in the development of silicosis. I. Biological characterisation of a granulocyte-activating mediator from human monocytes/ macrophages after treatment with quartzdust DQ-12. Ergebnisse von Untersuchungen auf dem Gebiet der Staub- und Silikosebekämpfung im Steinkohlenbergbau, Bd. 16, 1986, in press

5. Maly, F.-E. and A. L. de Weck : Activation of PMN-superoxide production by TNF, GM-CSF, and different mononuclear cell supernatants. Int. Archives of Allergy and Applied Immunology, Vol. 83, Suppl. 1, 1987, S. 30

6. Schmidt, J. A., C. N. Oliver, J. L. Lepe-Zuniga, I. Green and I. Gery: Silica-stimulated monocytes release fibroblast proliferation factors identical to interleukin 1. J. Clin. Invest., Vol. 73 , 1984, S.1462/1472

IN VITRO EFFECTS OF ENVIRONMENTAL POLLUTANTS ON MEDIATOR RELEASE

Rohr, U., W. König and F. Selenka
Institute of Hygiene and Medical Microbiology at the University of Bochum, Universitätsstr. 150, 4630 Bochum, FRG

INTRODUCTION

Interest has centered on the immune system as a target organ for environmental pollutants. In recent years it has been shown that many industrial chemicals and pesticides can influence various immunological functions. One component of the immune system are mast cells and basophile granulocytes. These cells are involved in all forms of allergic and inflammatory reactions. Mast cells and basophils can be stimulated to release mediators of inflammation. These mediators, like histamine, chemotactic factors, and leukotrienes, are responsible for the symptoms of allergic processes, e.g. swelling, redness, and bronchoconstriction. There are a variety of ways by which mast cells and basophils can be stimulated to degranulate. The most important immunological mechanism is the cell-bound IgE-antibody-antigen-reaction. Other immunological stimuli are the complement factors C3a, C4a, C5a, bacteria, and bacterial products. Furthermore the ability to release mediators in vitro and/or in vivo has been demonstrated for many non-immunological substances, like drugs, and food additives, e.g. radiocontrast material, benzoic acid, and Natriumsulphite. Our studies have shown the capability of environmental pollutants to release mediators from mast cells and human basophils.

MATERIALS AND METHODS

We worked with RPC (rat peritoneal exudate cells) containing 5 to 15 % mast cells. The cell suspensions were obtained from the peritoneal cavity of Wistar rats. Human basophils were isolated from the blood of healthy donors seperated on a Ficoll-metrizoate gradient. Basophils were concentrated in the fraction of lymphocytes-monocytes-basophils (LMB) which contain 0.1 to 1 % basophile granulocytes. The cells were stimulated with various concentrations of environmental pollutants. After incubation for 60 min cell supernatants were assayed for histamine or leukotrienes. The immunological control of our testsystem was functionally purified anaphylatoxin C5a which was isolated from human serum.

RESULTS

The results clearly show that various hazardous chemicals like organochlorine pesticides and chlorinated biphenyls induce a dose-dependent mediator release from rat mast cells and human basophils (Fig. 1). Incubation of RPC with dieldrin produces a dose-dependent histamine release in concentrations ranging from 0.01 to 10 µg/ml. A similar effect is observed after stimulation with DDT. Further studies have shown that not only preformed mediators but also newly generated lipid mediators can be released from RPC after stimulation with organochlorine pesticides. Dieldrin and DDT induce a dose-dependent release of leukotriene C4 parallel to the histamine release.

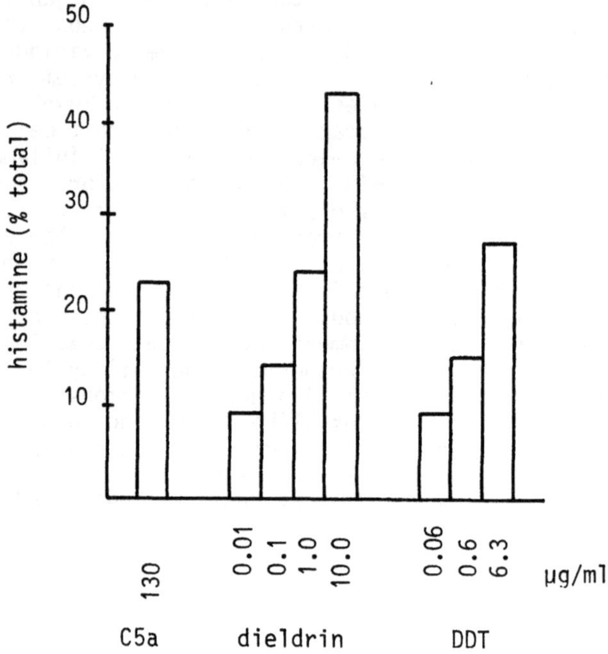

Fig. 1: Release of histamine (% of total histamine content minus buffer control) from RPC after incubation with organochlorine pesticides or C5a for 60 min.

Another group of substances with environmental significance are chlorinated biphenyls. Histamine release from human basophils was studied after stimulation with several PCB congeneres. In this case we tested only those substances which are the major components of the PCB residues in human blood and fat. The lower chlorinated substances show a similar pattern of histamine release whereas the sixfold chlorinated biphenyls produce a marked lower effect (Fig. 2). The histamine release is induced by PCB-concentrations from 4 to 8 µg/ml. It is of great interest to discuss these in vitro effects in relation to mans chronic PCB

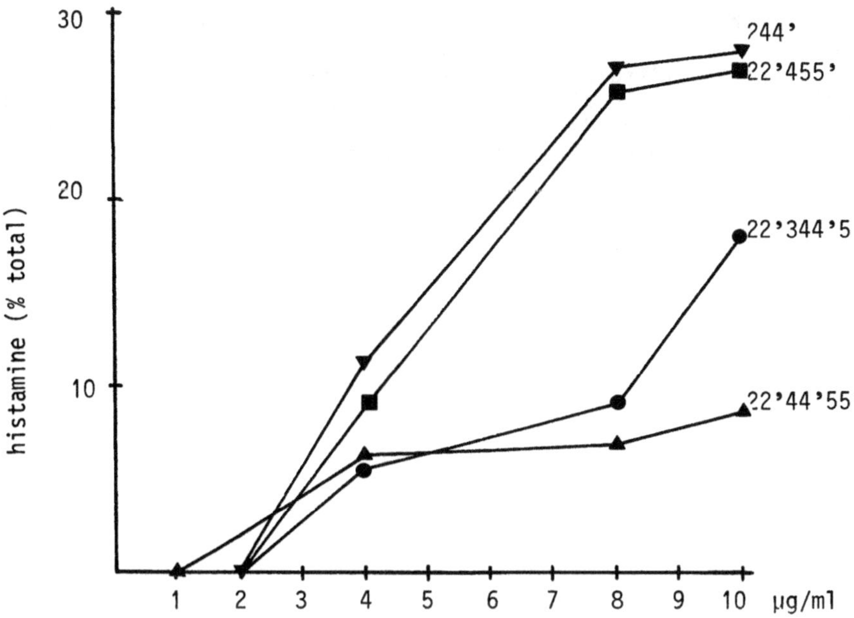

Fig. 2: Release of histamine (% of total histamine content minus buffer control) from LMB after incubation with chlorinated biphenyls for 60 min.

exposure. The concentrations of PCB congeneres in human blood range from 0.02 to 13 µg/l. We calculated a factor between PCB concentration needed to induce the effect and PCB concentration of human blood. This factor is 300 to 10.000. In view of the high PCB concentrations required to elicit the effect, interpretation of its physiological significance must await results from further studies.

A toxicological evaluation of the chemical-induced mediator release is only possible if combination effects are considered. Our studies clearly show that stimulation of human basophils with 1 to 3 environmental pollutants results in subadditive, additive, or synergistic effect on mediator release.

SUMMARY AND CONCLUSIONS

Our studies have shown that organochlorine pesticides and chlorinated biphenyls can stimulate mediator release in vitro acting by non-immunological mechanisms on mast cells and basophils. Further studies will clear up if these effects are important in vivo too. We expect that mediator release induced by hazardous chemicals may be one component of allergic or pseudoallergic processes in man. Together with immunological stimuli, like IgE, and other non-immunological factors, like drug additives, environmental pollutants might cause effects of pathophysiological relevance.

ALTERATIONS OF CHICK EMBRYO YOLK SAC BLOOD VESSELS AFTER APPLICATION OF ENVIRONMENTAL POLLUTANTS

Rosenbruch, M., A. Holst and W. Hilscher
Medical Institute of Environmental Hygiene at the University of Düsseldorf, Gurlittstr. 53, D-4000 Düsseldorf 1 (F.R.G.)

INTRODUCTION

Since during last years the amount of potential environmental pollutants, which have to be tested concerning their harmfulness, and the importance of animal protection are increasing more and more, appropriate test systems need to be found.
One of these systems to test the effects of harmful substances is the fertilized and incubated chicken egg. This system is used since many years in different fields, such as embryology, teratology etc. The following paper shows some methodological aspects how the chick embryo and the yolk sac blood vessel system can be used as test systems in environmental research.

MATERIAL AND METHODS

The tests are carried out on fertile White Leghorn eggs, obtained from Lohmann Tierzucht, Cuxhaven (F.R.G.). The test compounds can be administered via different routes: injection into the air chamber, the albumen or the yolk before or during the first days of development, or direct application on or into the veins of the yolk sac vessel system (Christ, 1973).
Before handling the eggs are cleaned with ethanol (70 %) and the injection is done through a drilled hole in the shell. Following the injection the hole is closed using commercial plastic tape.
Injection in the present study was done prior to incubation into egg white and yolk and the evaluation of findings took place from 3rd to 7th day of incubation.
For testing irritating potentials of substances the easiest way is direct application on the blood vessels. The substances are solved or suspended in a mineral solution and different concentrations and various amounts can be applicated on the vessels on day 3 or 4 of incubation (fig. 1).
After administration of the substances in tests using the direct application method and after the first evaluation at certain days in injection tests the eggs are covered with a paraffin shell including an observation window and are incubated again. In this way further development of the embryo, the blood vessel system and the alterations of the yolk sac blood vessel system can be observed up to 72 hours. Additionally surviving rates of the embryos at certain times after application can be detected.

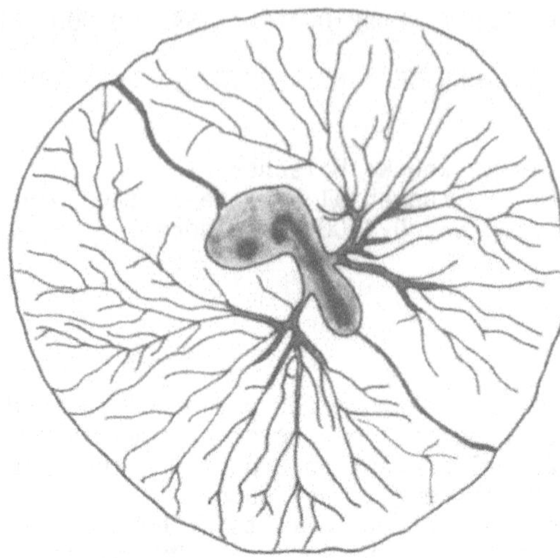

Figure 1 : Schematic depiction of chick embryo on day 4 of incubation surroundes by the yolk sac blood vessel system

RESULTS

- Injection study

In a study on heavy metal effects on the yolk sac blood vessels cadmium chloride was solved in a mineral solution and injected. Opening the eggs from day 3 up to day 7 of incubation, a distinctly increased rate of non developed chick embryos and an increased mortality occurs in the Cd-treated groups compared with the injection of the solvent and the untreated controls. Furthermore a mild difference is obvious between yolk groups and egg white groups; a higher mortality rate occurs in the yolk groups, after cadmium as well as after the solvent injection.

In surviving embryos a rarefication of the yolk sac blood vessels, embryonal and extraembryonal haemorrhages and additional vessel nets out of the sinus terminalis occur. These findings are possibly due to inhibitory effects of cadmium on the angioblasts.

- Direct application test

The test samples used here belong to a group of irritating substances, called "set - accelerating - admixtures". They are used in tunneling to accelerate the process of hardening of the shotcrete. Dropping 10 micro-liters supernatant of a suspension on the normal vessel system at day 4 of incubation, hyperaemia and capillary haemorrhages occur after about one minute and severe blood vessel ruptures after 3-5 minutes.

Different substances and different concentrations were tested concerning the "early-reaction" of the yolk sac blood vessels, within the first five minutes after application. Looking at the time when

certain findings were detectable, very distinct differences were observed between various concentrations of one substance and between three different test substances.
Three days after application a circumscribed reaction nodule, consisting of the haemorrhages in the center and newly developed reaction tissue in the periphery exists. In some cases this nodule is surrounded by a secondary blood vessel sinus.
This "late reaction" of the blood vessel system on the alterations can be studied from 24 hours post application up to 72 hours using different morphological as well as functional methods, such as light microscopy, electron microscopy and enzyme histochemistry.
In histology, different reaction processes are obvious: serosa like cells covering the damaged area, spindle-shaped mesenchymal cells organizing the necrosis, phagocyting cells with positive reaction on acid phosphatase and collagenous fibres as early as two days after application.

DISCUSSION

Chick embryos are used very often and since many years in medical research, in order to find teratological effects (Goerttler, 1962), to study developmental stages or to clarify biochemical processes and alterations. But mostly later stages of development and differences in hatching rates after exposures have been studied (Birge and Roberts, 1976).
In toxicology the chorio-allantoic-membrane (CAM), the second extra-embryonal blood vessel system, recently is used as a test system in different ways (Leighton, 1985; Lüpke, 1985).
But no results are available to ascertain if the chick embryos yolk sac blood vessel system is suitable for testing the effects of environmental pollutants.
The studies described here verify that different methods exist to use this system in the fields of experimental pathology and experimental toxicology. Using this modell it is possible to elucidate effect mechanims and to estimate irritating potentials of certain compounds. Satisfactory substance depending effects and good dose effect relations could be obtained in injection as well as in direct application tests.
Due to the fact that the onset of first reflexes and sensivity takes place about 6th to 7th day of incubation (Freeman and Vince, 1974), these studies are done on living and dynamically developing, but pain-free test samples.
Besides this, there are some other advantages compared with the above mentioned modells of the CAM : possibility of evaluating effects on the embryo after intravascular administration of pollutants, observing the whole system including the embryo during the experiment, easier handling and shorter incubation period before the test. Furthermore all morphological and functional methods, such as electron microscopy and enzyme histochemistry can be used (Rosenbruch, 1987).

CONCLUSIONS

It is possible to study the effects of environmental pollutants on chick embryos and the yolk sac blood vessel system during the first third of development.
In this way the total amount of experiments on living mammals can be reduced.
But it will not be possible to replace all animal experiments using so called alternative methods.

REFERENCES

Birge, W.J. and O.W. Roberts (1976) : Toxicity of metals to chick embryos. Bull.Environm.Contam.Toxicol. 16, 319-324

Christ, B., H.J. Jacob and M. Jacob (1973) : Über ein Verfahren zur Gefäßdarstellung beim Hühnerembryo. Der Präparator 19

Freeman, B.M. and M.A. Vince (1974) : Development of the avian embryo. Chapman and Hall Ltd., London

Goerttler, K. (1962) : Der teratologische Grundversuch am bebrüteten Hühnerkeim, seine Möglichkeiten und Grenzen. Klin.Wschr. 40, 809-812

Leighton, J., J. Nassauer and R. Tchao (1985) : The chick embryo in toxicology: an alternative to the rabbit eye. Food Chem.Toxicol. 23, 293-298

Lüpke, N.P. (1985) : Hen's egg chorioallantoic membrane test for irritation potential. Food Chem.Toxicol. 23, 287-291

Rosenbruch, M. (1987) : Alterationen und Reaktionen des aviären Dottersackgefäß-Systems nach direkter Schadstoffapplikation. Verh.Dtsch.Ges.Pathol. 71, 438

USE OF BOVINE ALVEOLAR MACROPHAGES AND OF THEIR LYSOSOMES FOR IN VITRO STUDIES WITH MINERAL DUSTS

Seidel, A., E. Drosselmeyer, G. Hotz, S. Pätzold, J. Schimmelpfeng, B. Walser
Kernforschungszentrum Karlsruhe, Institut für Genetik und Toxikologie, Postfach 3640, D-7500 Karlsruhe 1, FRG

INTRODUCTION

The alveolar macrophage and its lysosomal system play an important role in many physiological and toxicological processes. The exocytosis or intracellular release of lysosomal material probably stand at the beginning or are accompanying a sequence of pathogenetic events leading to cell death and lung injury. For mechanistic studies with cell organelles as well as for repeated studies with genetically homogeneous cells it is of great advantage to work with large numbers of cells (10^9 or more). In the case of cells from the monocyte-macrophage system such large numbers can often not be obtained. Desai and Richards (1) have used sheep alveolar macrophages recovered by post-mortem lavage for studies with asbestos. In view of the easiness to obtain bovine lungs, which contain still larger numbers of macrophages, we have started to characterize bovine alveolar macrophages (BAM), their cell organelles and their reaction to particles like Latex, TiO_2 or quartz taking advantage of the experiences of previous authors (e.g. 2) with these cells.

MATERIALS AND METHODS

BAM were obtained within one hour after death preferably from healthy young males by lavage with about 18-20 l 0.9% NaCl and first centrifugation steps followed by centrifugation in Ficoll-Paque in order to eliminate dead and non-macrophage cells. Depending on the chemical nature of the plastic vessel, suspension or monolayer culture can be performed using routinely available media (usually RPMI1640) and Penicillin/Streptomycin.Vitality was assayed by Trypan Blue. Light- and electron microscopy and staining were performed with routine procedures, exposure to non-radioactive particles in monolayer culture. Further details and references regarding methods can be found in (3) and (4).

RESULTS

By post-mortem lavage we were able to obtain routinely up to 10^9 viable alveolar macrophages with a purity of >90%. They can be kept in culture for at least 48 hours without problems, though lavage was not performed under sterile conditions. The cells are positive for unspecific esterase, show a very high phagocytic activity with Latex and express Fc receptors (see also 2). Upon stimulation with opsonized Zymosan luminol-dependent chemoluminescence occurs.
The results of the analysis of the mitochondria-lysosome fraction by carrier-free electrophoresis are presented in Fig.1. Lysosomes (NAG and SP) have a higher negative charge and are well separated from the mitochondrial peak (GDH) which also represents the position of endoplasmatic reticulum and pericellular membranes (not shown). By electron micro-

scopic examination of the pure lysosomal fraction various types of lysosomes could be discerned, further ultrahistochemical characterization is under way. These lysosomes can be kept stable in suitable media for at least 90 min.
BAM are relatively resistent to temperature changes during in vitro cultivation and to cryopreservation. Repeated changes of temperature from 37 to 4 C caused some loss of vitality but the bulk of the cells remained viable. After cryopreservation about 50% of the cells were viable. As judged from their adherence behaviour and from their phagocytotic activity they were also functionally intact, at least during the first hours after thawing.
The cells were successfully used to study the uptake and intracellular fate of respirable nuclear fuel particles (for details see 4) and their response towards non-radioactive particles was analysed. Non-fibrogenic particles like Latex and TiO_2 caused release of lysosomal enzymes but no cytotoxicity. In contrast DQ_{12} quartz caused immediate or delayed cell death with release of lactate dehydrogenase and N-acetylglucosaminidase. This effect was distinctly delayed when surfactant from the same animal was present in the culture medium (Fig.2).

CONCLUSIONS

Bovine alveolar macrophages can be obtained with high yield and purity by post mortem lavage, they show all the characteristics of macrophages from other animal species. Their reaction towards dust particles is qualitatively and quantitatively similar to that of other types of alveolar macrophages. The very large number of cells makes it possible to repeat serial studies with genetically homogeneous BAM.

ACKNOWLEDGEMENTS

The authors are gratefull to Prof. Seemayer for supplying DQ_{12} quartz and to Dr. Baader and M. Wiener for performing the chemoluminescence assays.

REFERENCES

(1) Desai, R., R.J. Richards (1983) Effects of Chrysotile on a lysosomal enzyme preparation and on the hydrolytic enzyme activity of cultured alveolar macrophages, Environm. Health Perspectives, 51, 125-130.

(2) Bielefeldt Ohmann, H., L.A. Babiuk (1986) Bovine alveolar macrophages: Phenotypic and functional properties of subpopulations obtained by Percoll density gradient centrifugation, J. Leucocyte Biol., 39, 167-181.

(3) Walser, B., G. Hotz, H.L. Müller, S. Pätzold, M. Wiener, A. Seidel (1987) Untersuchungen an Rinderalveolarmakrophagen und deren Organellen vor und nach Cryopreservation, Report des Kernforschungszentrums Karlsruhe, KFK-PEF, 12, 685-695.

(4) Müller, H.L., E. Drosselmeyer, G. Hotz, A. Seidel, H. Thiele, S. Pickering, I.L. Ray (1987) Zum biologischen Verhalten von Uran-Plutonium-Mischoxidaerosolen: Inhalationsversuche mit Ratten und in vitro-Studien mit Alveolarmakrophagen, Report des Kernforschungszentrums Karlsruhe und Euratom, KFK 4172 / EUR 10846 DE.

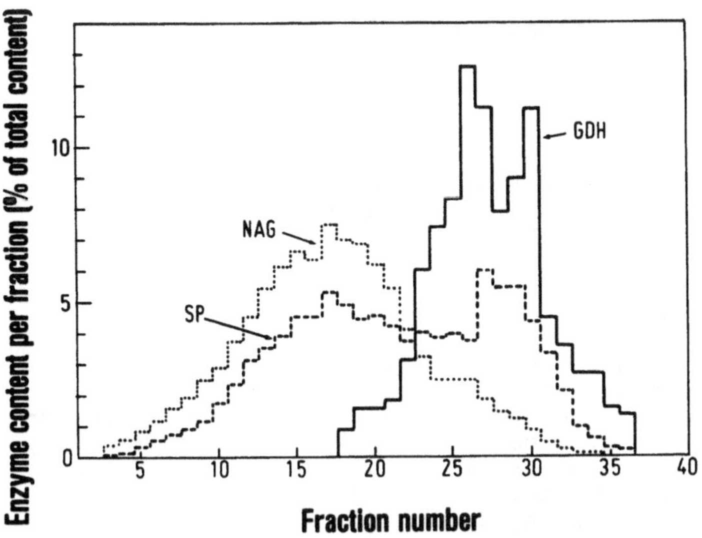

Fig.1: Marker enzymes after carrier-free electrophoresis of the mitochondria-lysosome fraction from BAM (Anode left side). NAG: N-Acetylglucosaminidase; SP: Acid Phosphatase (both lysosomes); GDH: Glutamatedehydrogenase (mitochondria).

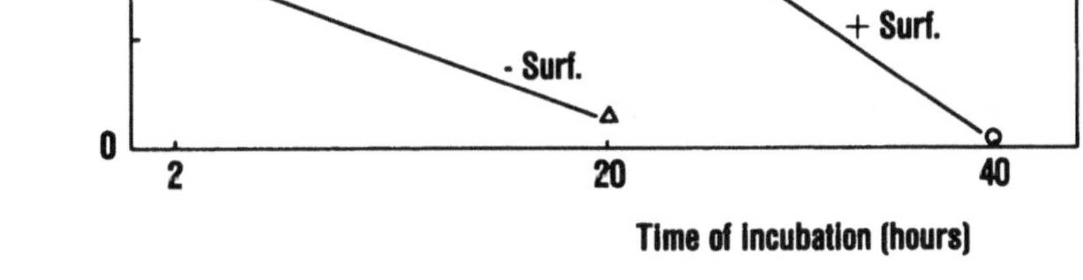

Fig.2: Vitality of BAM after incubation with TiO_2 (● Anatase) and DQ_{12} quartz (both ∿ 1 µm diameter, 500 µg/ml). ± Surf.: quartz with or without surfactant from the same animal.

Neurotoxicology of heavy metals: Synaptic transmission as influenced by mono- and divalent metalcations.

Wiegand, H.
Medical Institute of Environmental Hygiene at the University of Düsseldorf, Auf`m Hennekamp 50, D-4000 Düsseldorf, FRG

Introduction

Divalent, essential cations like magnesium-, zink- or calciumions play a pivotal role in many biological processes and are effective in relatively low concentrations. In many organs the action of heavy metals is manifested by an interaction with the physiological functions especially of calcium, i.e. in the nervous system the regulation of transmitter-release in synaptic transmission.
The heavy metals thallium and lead belong to different groups of the periodic system and therefore they are not very similar to each other. As to their action onto synaptic transmission in the nervous system, there are - at least at the first glance - asthonishing similarities.

Materials and Methods

Synaptic transmission was studied in the neuromuscular synapse, the anatomy and physiology of which is well known (1) and which is thought to be the typical model synapse. Intracellular microelectrode-recordings from muscle cells and extracellular focal recording from presynaptic terminals were used, respectively (2, 3). The preparations were exposed to thallium via the Ringer solution.

Results

Thallium was capable to interfere with synaptic transmission (Fig. 1).

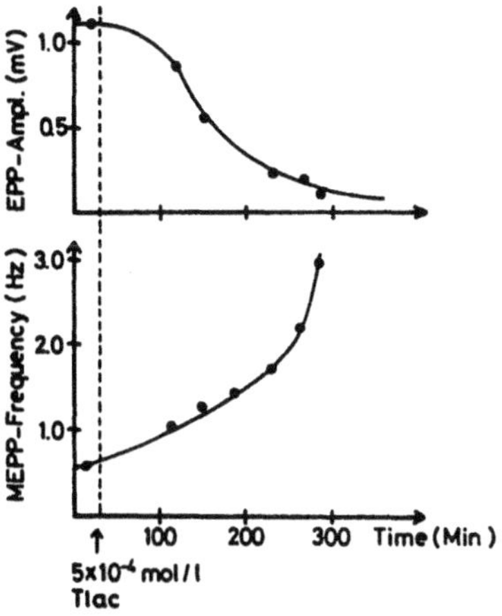

a Panel a: Decrease of evoked synaptic activity upon thallium-acetate-superfusion (Tlac) of a neuromuscular junction.

b Panel b: Enhancement of spontaneous synaptic activity upon thalliumacetate-superfusion (Tlac) of a neuromuscular junction.

Fig. 1

It induced enhancement of spontaneous transmitter release, whereas evoked transmitter release was blocked. Just the same phenomena were recently described by Cooper (4) for the action of lead onto synaptic transmission in the neuromuscular junction. In contrast to Cooper`s findings for lead-ions the blockade by thallium of evoked transmitter release was not reversible. Cooper (4) gave several evidences, that lead blocked synaptic transmission by antagonizing praesynaptically the influx of calcium-ions via voltage regulated ion-channels of the nerve terminals. It might be speculated that in analogy to the divalent metal-ion lead thallium might have blocked calcium-influx into presynaptic nerve terminals. Though such a mechanism was largely improbable because of irreversibility of thallium action as compared to the reversible lead action, it was tempting to measure presynaptic calcium-currents using focal, extracellular recording from presynaptic terminals (see 5).
Using this experimental technique we obtained presynaptic current signals under special conditions, which are proved to be caused by calcium ion influx (6). These signals could be blocked by application of the divalent metal-ion cobalt, a well known inorganic calcium-channel-blocker (Fig. 2).

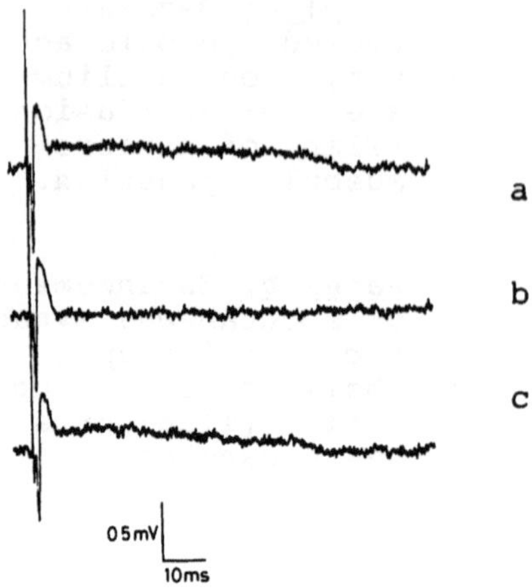

Ca^{++}-current signal before (a), 10 Min after Ca^{++}-superfusion ($5x10^{-5}$M) (b) and 10 Min after washing with normal Ringer`s solution (C).

Fig. 2

As to the effects of thallium-ions onto such presynaptic calcium currents we never got any evidence for it to antagonize these calcium-ion currents in presynaptic nerve terminals even using extremely high thallium concentrations up to 1 mM (Fig. 3).

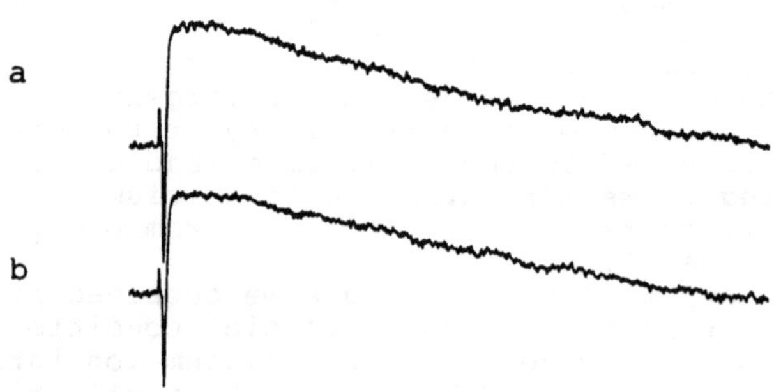

Ca^{++}-current signal before (a) and 30 Min after Tlac-superfusion (1 mM)

Fig. 3

Discussion

Synaptic transmission in the neuromuscular junction is achieved by the action potential invading the presynaptic nerve terminal and releasing acetylcholine (Ach). Ach diffuses through the synaptic cleft and reaches the postsynaptic membrane receptors to induce endplate potentials. This is stimulus-evoked or phasic transmitter release, besides which there is also spontaneous Ach-release, which is seen as miniature endplate potentials (MEPP's) - the electrical correlate of randomized release of synaptic vesicles. Electrophysiological analysis of MEPP`s-amplitudes and -frequencies allowes to differentiate between pre- and postsynaptic effects of heavy metals. Interaction with presynaptic transmitter release would induce variation of MEPP-frequencies exclusively, which is indeed found for thallium. In contrast action on postsynaptic transmitter receptors would induce changes in MEPP-amplitudes, a phenomenon which was never found for thallium. However, there were important differences as to the action of lead and thallium onto presynaptic phenomena. Whereas Cooper (4) described a competetive antagonism to calcium-ions for lead in regulation of Ach release, the monovalent thallium-ion failed to antagonize presynaptic calcium influx. So far, there seemed to be the following differences as to their mode of action:

1. Lead antagonized calcium ion function in mediation of presynaptic Ach release presumably by blocking calcium-ion-chanels of the presynaptic nerve-membrane from the outside; this action was reversible.

2. Thallium obviously did not interfere with presynaptic calcium ion influx, but nevertheless blocked phasic transmitter release; this action was irreversibel. It might be therefore speculated, that thallium has to be transported into the intracellular space of presynaptic nerve terminals to interfere with intracellular organelles necessary for Ach release by regulation of presynaptic concentration of free calcium ions.

Concluding, there are some evidences for an antagonistic interfering of metallions with calcium regulating processes of the presynaptic nerve endings during synaptic transmission. However, these evidences are far from drawing a conclusive picture of how metal-ions act.

References

(1) Katz, B. (1969): The release of neural transmitter substances. Liverpool University Press, Liverpool, U.K.

(2) Wiegand, H., H. Lohmann and S.V. Chandra (1986): The action of thallium acetate on phasic transmitter release in the mouse neuromuscular junction. Arch. Toxicol., 58, 265-270

(3) Wiegand, H., H. Lohmann, S. Uhlig 81986): The action of thallium, cadmium and cobalt on presynaptic Ca^{++}-currents in mouse motor nerve terminals. Naunyn-Schmiedeberg`s Arch. Pharmacol. 334 (Suppl.), R16.

(4) Copper, G.P. and R.S. Manalis (1983): Influence of heavy metals on synaptic transmission: A review. Neurotoxicology 4, 69-84.

(5) McArdle, J.J., D. Angoit-Petit, A. Mallart, L. Faille and J.L. Brigant (1981): Advantages of the triangularis sterni muscle of the mouse for investigations of synaptic phenomena. J. Neurosci. Meth., 4, 109-115.

(6) Penner, R. and F. Dreyer (1986): Two different presynaptic Ca-currents in mouse motor nerve terminals. Pflügers Arch, 406, 190-197.

II. ANIMAL MODELS

EFFECT OF SOME METAL IONS (Cd^{++}, Pb^{++}, Mn^{++}) ON MEDIATOR RELEASE FROM MAST CELLS IN VIVO AND IN VITRO

Behrendt, H., M. Wieczorek, S. Wellner and A. Winzer
Department of Electronmicroscopy, Medical Institute of Environmental Hygiene at the University of Düsseldorf, Auf'm Hennekamp 50, D-4000 Düsseldorf, FRG

INTRODUCTION

Mast cells are tissue bound mononuclear granulated cells deriving from the pluripotential hematopoietic stem cell (1). The cells can be easily identified by the strong affinity of their granules to basic dyes, i.e. toluidine blue and alcian blue at low pH. By electronmicroscopy mast cells exhibit specific granules which are characterized by a special, species-dependent matrix substructure. Mast cells are located at the border of inner and outer surfaces within mammalian organisms, and they are said to play an important role in maintaining physiological microenvironmental conditions via regulating vascular permeability through histamine and other mediators.

Under pathological conditions mast cells are inflammatory cells. They are involved in a variety of human diseases, i. e. fibrosis, tumorigenesis, chronic bronchitis and allergy. In immediate allergic inflammatory reactions mast cells display a central role because of the presence of high affinity IgE-Fc receptors on their cell surface as well as of the synthesis and generation of mediators of allergic reactions.

In the course of our investigations on the effects of environmental pollutants on IgE-mediated allergic reactions, the influence of lead, cadmium and manganese ions on mediator release from mast cells has been studied. The heavy metals cadmium and lead are overall distributed and are known to exhibit neurotoxic, nephrotoxic and immunosuppressive acitivity. Manganese compounds are generally said to promote respiratory disease in man and laboratory animals (2). Manganese dioxide potentiates experimental respiratory infections in mice (3). Moreover, the regulation of thrombin-induced degranulation of mucosal mast cells by manganese has been reported (4).

The present study deals with the stimulus-induced histamine release from rat peritoneal mast cells which have been exposed to nontoxic concentrations of metal ions either in vivo or in vitro.

MATERIALS AND METHODS

In vivo experiments.
20 female Wistar rats, 8 weeks old, were fed with a standart diet containing either 250 ppm lead (Pb-acetate) or 50 ppm cadmium (Cd-chloride). After an exposure period of

6 months the animals were killed by decapitation and the peritoneal washout cells were harvested according to the method of Uvnäs and Thon (5). In order to evaluate animal burden, blood levels of cadmium and lead were determined by atomic adsorption spectroscopy. At the time of killing the cadmium level per 100 ml blood was 25 $\pm$ 2.2 µg, the blood lead level was 17 $\pm$ 3.8 µg/ml, resp.

In vitro experiments.
Animals. Male, 10-12 weeks old Lewis rats were used throughout these experiments.

Pollutants. The cationic metal compounds lead acetate, lead chloride, cadmium chloride and manganese chloride were used in this study.

Secretagogues. For induction of histamine release from mast cells the stimuli calcium ionophore A23187 and Compound 48/80 have been employed.

Assay for histamine. Histamine was determined fluorimetrically by using the Technicon Autoanalyzer.

Electron microscopy. For ultrastructural analysis the cells were fixed with glutaraldehyde and postfixed with osmiumtetroxide. After dehydration the samples were embedded in plastic resin. Ultrathin sections were investigated in a Philips 400T electron microscope.

Course of experiments. Peritoneal washout cells were counted and the percentage of mast cells was determined. In addition cell viability tests using Eosin Y have been performed.

For all experiments triplicate samples of 5 x 10^4 mast cells suspended in 0.4 ml Hanks balanced salt solution containing Ca2+ and Mg2+ (HBBS) were incubated at 37°C in a gently shaking water bath for 5 min. Then, 0.1 ml HBBS containing incremental concentrations of metal cations (0.001-1000 µM) were added to the samples and the cells were incubated for another 10 or 30 min at 37°C in order to determine the influence of these compounds on histamine release. Alternately cells treated with one of the metal cation for 30 min were challenged with either ionophore A23187 (0.1 µg/ml) or compound 48/80 (o.1 and 0.5 µg/ml) for another 10 min. In other sets of experiments triplicate samples of 5 x 10^4 mast cells were incubated for 10 min at 37°C in HBBS containing both, incremental concentrations of the cation and of the stimulus used. After the incubation period all cells were sedimented at 1000 rpm for 10 min at 4°C and the pellets and supernatants were assayed for histamine.
Each experiment was repeated at least two times. Histamine release was expressed as percent of the total histamine content. In order to compare the effects of various metal ions on the release process, a concentration which induces a release of about 40% of the total histamine has been chosen for the secretagogues. The results obtained are then expressed in percent of the histamine release induced by the stimulus alone.

Statistical analysis has been performed by Student t-test.

RESULTS AND DISCUSSION

The results obtained from Wistar rats which had been chronically exposed to cadmium and lead are shown in Table 1. As can be seen, there was a reduction of the histamine content per single mast cell in metal-treated animals. Additionally, spontaneous histamine release increased up to one third of the total histamine content, whereas the histamine release induced by ionophore A23187 was inhibited in both, the cadmium-treated and the lead-treated group.

From these pilot studies it can be concluded that the metal ions cadmium and lead do have some effect on rat peritoneal mast cells *in vivo* in influencing both, mediator content and mediator release.

We further analyzed as to whether the same effects occured under *in vitro* conditions. For this purpose mast cells were isolated from the peritoneal cavity and subsequently incubated with cadmium, lead and manganese. The results obtained indicate that neither cadmium nor lead nor manganese had any histamine liberating activity by itself. Therefore, it has been tested as to whether these metal ions were able to influence the stimulus-induced histamine release from rat peritoneal mast cells. The results obtained from these experiments are shown in figure 1. As far as lead is concerned, it demonstrates that there was no effect at all, neither by simultaneous application of the stimulus nor after preincubation of the cells with lead and subsequent stimulation. Additionally, stimulus-induced histamine release was not affected by physiological concentrations of manganese. Only at higher concentrations a slight inhibitory activity could be observed (fig. 1).

Table 1. Effect of longterm exposure of rats to cadmium and lead on histamine content of peritoneal mast cells and on the ionophore A23187 induced histamine release. (n=10). From (6).

	control	cadmium^{+}	lead^{++}
histamine content (pg per mast cell)	22.8	8.5	6.7
% histamine release			
spontaneously	4.7	10.5	33.6
ionophore 0.1 µg/ml	50.3	17.5	23.5

$^{+}$) diet containing 50 ppm

++) diet containing 250 ppm

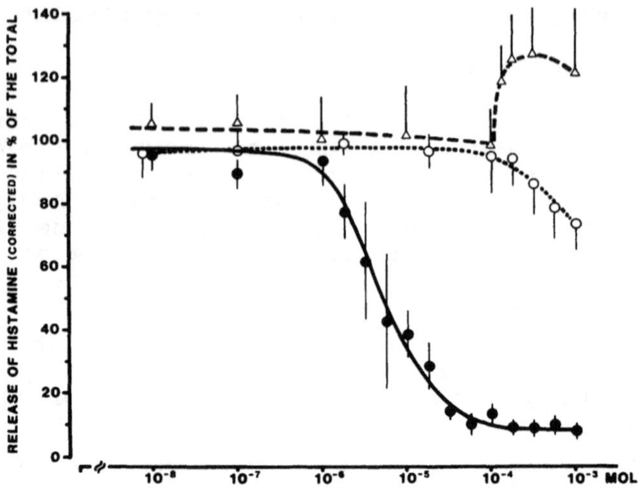

Figure 1. Effect of cadmium (●——●) and lead (Δ ---Δ) on the Ca-ionophore A23187-induced and of manganese (o···o) on the compound 48/80-induced histamine release from rat peritoneal mast cells in vitro. Each value represents mean ± SD, n = 9. Modified from (6).

Since lead as well as manganese cations had no direct effect on histamine release in vitro, the results obtained from the in vivo experiments are not restricted to mast cells and may be due to an impairment of the microenvironment by lead.

In contrast to lead and manganese, cadmium chloride inhibited histamine release induced by ionophore A23187 or by compound 48780 in a dose-dependent manner (figure 1). This inhibitory action of cadmium on mast cell activation was only partially due to a calcium antagonistic effect (6). Therefore, binding sites for cadmium, but not for lead and manganese, probably exist in the plasma membranes of mast cells, as is true to occur in nerve cell membranes.

Since after preincubation of the cells with cadmium and subsequent stimulation the inhibitory effect persisted and could not be abolished by washing the cells prior to stimulation, we concluded that cadmium had been bound to the cells. This assumption was confirmed by experiments that demonstrated a recovery of histamine release when freshly isolated mast cells were stimulated in the presence of a cadmium-containing supernatant from another inhibition experiment.

In a further set of experiments we investigated cadmium-treated and stimulated mast cells by transmission electron microscopy. Figure 2 demonstrates an electron micrograph of a part of a degranulating mast cell in the presence of low concentrations of cadmium. It indicates that low concentrations of cadmium inhibited the dissolution of granular ma-

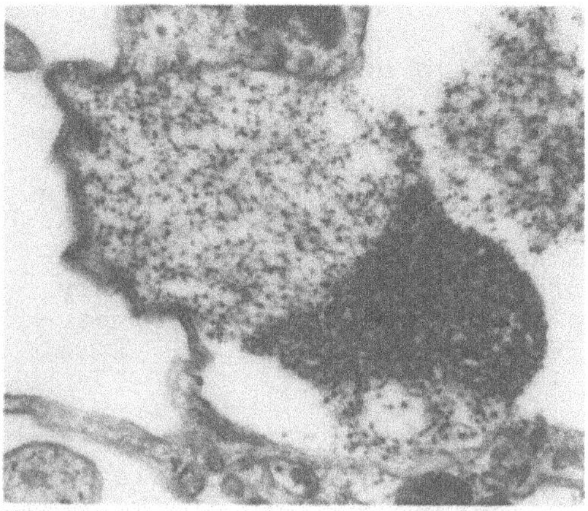

Figure 2. Incomplete dissolution of the granular matrix from degranulating mast cells by cadmium (5 x 10^{-6}M). Electron micrograph, magnification 37ooo : 1

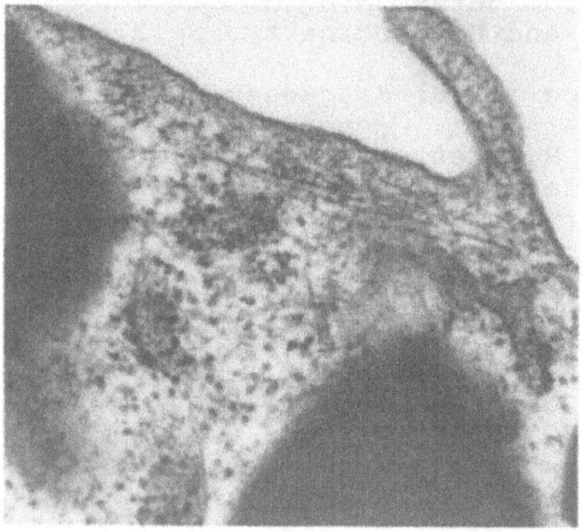

Figure 3. Induction of microfilaments in the periphery of stimulated mast cells by cadmium (10^{-4}M). From (6). Electron micrograph, magnification 54ooo : 1

trix material which already had been exposed to the extracellular space during degranulation. In terms of mediator release it can be said that there is a cadmium-induced dissociation of the release between preformed mediators, i.e. histamine, and mediators i.e. leukotrienes generated in the presence of a stimulus.

The effect of high concentrations of cadmium on stimulus-induced mast cell degranulation resulted in total inhibition of secretion. Ultrastructurally, this effect was accompanied by large amounts of subplasmalemmal microfilaments and microtubules (figure 3). Therefore, cadmium ions (but not lead or manganese) stabilize cell membranes via activation of the transmembraneous cytoskeleton which in turn inhibits receptor-mediated movement of integral proteins in plasma membranes.

Since cadmium inhibits antigen-induced histamine release from human basophils (7), too, and also affects endothelial cells (8), an influence of cadmium on immediate type allergic reactions has to be expected.

SUMMARY

The effects of cadmium andlead _in vivo_ on mast cells as well as the influence of cadmium, lead and manganese on histamine release from rat peritoneal mast cells _in vitro_ has been investigated. Both, lead and cadmium affected mast cell parameters _in vivo_, but only cadmium inhibited stimulus-induced histamine release _in vitro_. The inhibitory action of cadmium was accompanied by the activation of the transmembraneous cytoskeleton. Therefore, cadmium seems to mainly act through receptors in the plasma membrane and only partially by an calcium antagonistic effect.

REFERENCES

(1) Schrader, J.W. (1986) The Ontogeny and Differentiation of Mast cells. A Review. In: Befus, A.D. et al (Eds) Mast Cell Differentiation and Heterogeneity, pp 27-43, Raven Press New York
(2) National Academy of Science (1973) Manganese. NAS Printing Office, Wash. DC, USA
(3) Maigetter, R.Z., Ehrlich, R., Fenters, J.D., Gardner, D. E. (1976) Potentiating Effects of Manganese Dioxide on Experimental Respiratory Infections. Environm.Res. 11, 386-391
(4) Baranes, D., Liu, T.F., Marx, G., Shalit, M., Razin, E. (1986) Regulation of Thrombin-induced Mast Cell Degranulation by Zinc and Manganese. Immun. Letters 12, 95-99
(5) Uvnäs, B., Thon, I.L. (1959) Isolation of 'biologically intact' mast cells. Exp. Cell Res. 18, 512-520
(6) Wieczorek, M., Behrendt, H. (1988) Wirkung von Cadmium und Blei auf Mediatorzellen allergischer Reaktionen. Allergologie, in press
(7) Vervloet, D., Okazaki, R., Ilea, V.S., Reisman, R.E., Arbesman, C.E. (1975) Histamine release from human leukocytes: Modulation by cadmium ions. J.Allergy Clin. Immunol. 56, 262-272
(8) Nolan, C.V., Shaikh, Z.A. (1986) The vascular endothelium as a target tissue in acute cadmium toxicity. Life Sci. 39, 1403-1409

INVESTIGATIONS OF THE LUNG CARCINOGENIC POTENTIALS OF SODIUM DICHROMATE AND CR VI/III OXIDE AEROSOLS IN WISTAR RATS

Glaser, U., D. Hochrainer and H. Oldiges

INTRODUCTION

Cr is not yet a hazard for the general public, but various epidemiological reviews were in the last decades definetively indicative of excessive lung cancer rates among workers exposed in the chromate-producing industry. The carcinogenic risk at several other working places with Cr VI is also suspected (Ref. 1-3).

It is generally accepted that there were excessive concentrations of slightly soluble Cr VI, e.g. Ca-chromate, in the air of old chromate producing plants, until the role of Cr in lung cancer was established. Among employes of that time an excessive risk up to 29 fold was reported (Ref. 4). However, most of these studies failed in informations on tobacco smoking history and exposure to other chemicals and radiation. Thus, it is possible to estimate how many cases resulted from Cr.

Meanwhile occupational hygiene and new technical measures have diminished the air concentrations and now the predominant contaminant at chromate-production is the water soluble sodium dichromate (Ref. 5). But nevertheless there is an urgent need to determine the risk for workers today.

Our detailed knowledge on the carcinogenicity of Cr originates from experimental animal results. From the results of such studies (Ref. 6, 7) was derived that the low water soluble Cr-chromate, this is a mixture of sintered Cr VI/III oxide, and further Zn-, Ca-, and Sr-chromate are carcinogenic compounds, while for the water soluble Na-chromate the results indicate a questionable carcinogenicity. The knwoledge is still established in the MAC-value lists.

The results of Hueper & Payne (Ref. 6, 7) were confirmed by the results of Levy & Venitt (1975, Ref. 8), who did not find any lung cancer due to implants containing Na-dichromate. However, Steinhoff et al. (1984, Ref. 9) recently published a positive finding, when Na-dichromate was given intratracheally once weekly in the maximum tolerable dose of 1.25 mg/kg b.w. for lifetime. But the lung tumor incidence was as low as was found in the only existing long-term inhalation study with chromates that was performed with Ca-chromate and mice by Nettesheim et al. (1971, Ref. 10).

MATERIAL AND METHODS

This study considers the threshold values for air Cr of today. A long-term inhalation study with Wistar rats exposed to Na-dichromate and Cr VI/III oxide aerosols was performed using the same experimental design and equipment that succeeded in finding the carcinogenicity of Cd aerosols as already presented by Oldiges and Glaser (Ref. 11).

Groups of 20 male Wistar rats were continuously exposed for 22 hours a day and 7 days a week for 18 months in horizontal flow inhalation chambers. The Na-dichromate aerosol was generated by atomization of aqueous Na-dichromate solutions and diluted with filtered fresh air to the Cr concentrations of 25, 50 and 100 $\mu g/m^3$. The 2/3 mixture of the Cr (III/VI) oxide aerosol was obtained after partial reduction of Cr (VI) oxide solutions by hydrogen peroxide. Thereafter the nebulized Cr oxide aerosol was pyrolized at 200°C in a tube furnace and diluted to the Cr concentration of 100 $\mu g/m^3$. The mass median diameters of both Cr aerosols were 0.36 and 0.39 µm respectively. After the inhalation period of 18 months the rats were housed conventionally in single cages for a further year. One group of 40 rats was held as controls in filtered fresh air under the same conditions.

RESULTS

Figure 1, where the body weight gains for the Cr exposed groups and the controls are shown, gives an impression for that the long-term inhalation period of 18 months and also in the subsequent observation period of a further year the Cr exposures by both Cr aerosols were very well tolerated by the rats. There were also no differences in the appearance and behaviour of the rats or for the food and water consumptions between the groups. Further there were no findings from routine clinical chemistry and hematology except of slightly, but significant increased RBC counts, hematocrit values, hemoglobin levels, total WBC counts and elevated cholesterol levels in the blood for the Cr oxide exposed group at the end of the study, mostly due to moribund animals in this group. The mean tissue weights measured were just comparable between the rats of the experimental and control groups that survived the termination of the study in the 30th month. But significantly increased mean lung weights were determined also only for the Cr oxide group exposed to 100 $\mu g\ Cr/m^3$.

The retention of Cr in the lung and kidneys one year after termination of the inhalation is shown by the figure 2. Cr was measured in wet digested portions of the tissues by flameless AAS. The mean Cr contents in the lungs of the Na-dichromate groups exposed to 25, 50, and 100 $\mu g/m^3$ of Cr were 6, 13, and 50 µg Cr per organ respectively, while with about 500 µg Cr per organ there was much higher Cr retention in the lungs of the Cr oxide group exposed also to 100 ug/m^3 higher retention in the lungs of the Cr oxide group exposed also to 100 $\mu g/m^3$ of Cr. This difference may derive from the lower bioavailability of the Cr oxide mixture in the lungs of the rats compared with Na-dichromate. But surprisingly there were the same Cr retentions in the kidneys after exposure to both Cr aerosols.

The cumulative mortality incidences in the experimental groups and the controls are shown by the figure 3. During the inhalation period the mortality rates were very low among these groups. The mortality rates, however, increased with the beginning second year of the study, but were just comparable among the groups of the study.

In tissues other than the lungs Cr related lesions were neigther found macroscopically nor histopathologically. In the lungs of the controls no pigmentloaded macrophages were seen, while there were weak accumulations of pigmentloaded macrophages in the alveolar region of several rats of the lowest Na-chromate exposure group and moderate accumulations in the alveolar and peribronchial regions of all rats of the higher Cr exposed groups. Only for few rats of the Cr oxide mixture group there were also found eosinophilic substances inside the alveolar lumens combined with focal thickenend septa and interstitial fibrosis.

Primary lung tumors were not determined in the controls and also in the Na-dichromate groups exposed to 25 and 50 $\mu g/m^3$ Cr. However, three lung tumors, 1 adenocarcinoma and 2 adenomas, were evident in the Na-dichromate group exposed to 100 $\mu g/m^3$ Cr. In this group there was a further rat with a squamous cell carcinoma in the pharynx/larynx region. In the Cr oxide mixture group, also exposed to 100 $\mu g/m^3$ Cr, another rat was found with a primary lung adenoma. The tumors in tissues other than the lung could not be related to the Cr exposures.

CONCLUSIONS

These results with 20 % and 5 % of primary tumors in the respiratory tract of rats exposed to Cr aerosol levels of 100 $\mu g/m^3$ either of Na-dichromate or the Cr

(VI/III) oxide mixture indicate a weak, but real carcinogenic potency of both Cr aerosols. It is also obvious from these results that in this study there was not seen any toxicity except of slight findings in the lungs of the rats exposed to the Cr oxide mixture aerosol. Compared to these results Steinhoff et al. (Ref. 9) found nearly the same rat ot primary lung tumors after exposure to Na-dichromate instillations, but they also reported on moderate signs of toxicity. A much higher Cr aerosol concentration than in our study was used in the long-term inhalation study performed with Ca-chromate by Nettesheim et al. (Ref. 10). They exposed mice for 5 hours a day on 5 days a week to 13 mg/m^3 of Ca-chromate. But at nearly the same weekly dose of Cr they also found only a weak carcinogenicity as was seen in our study with Na-dichromate. And again due to the higher Cr aerosol level there were severe signs of toxicity in the mice. In conclusion, from these studies different soluble Cr compounds should be considered as weak carcinogens.

REFERENCES

(1) Hayes, R.B., in A.H. Lilienfeld (Edit.). Reviews in Cancer Epidemiology, Vol. 1, Elsevier/North Holland, pp. 293-333 (1980).

(2) Sunderman, Jr., F.W., Prev. Med. 5, 279 (1976).

(3) Langard, S., in D. Burrows (Edit), Chromium: Metabolism and Toxicity, CRC Press Inc., Boca Raton, Florida, pp. 13-33 (1983).

(4) US-Public Health Service, Health of workers in chromate producing industry, US Department of Health, Education and Welfare Publ. 192 (1953).

(5) Korallus, U. and N. Loenhoff, Arbeitsmed. Sozialmed. Preventivmed., 12, 285 (1981).

(6) Hueper, W.C. and W.W. Payne, Am. Ind. Hyg. J. 20, 274 (1959).

(7) Hueper, W.C. and W.W. Payne, Arch. Environm. Health, 5, 51 (1962).

(8) Levy, L.S. and S. Venitt, Brit. J. Cancer, 32, 254 (1975).

(9) Steinhoff, D., S.C. Gad, G.K. Gatfield and U. Mohr, Report to the Industrial Health Foundation, Wuppertal, F.R.G., Bayer AG., Inst. Toxicol. (1984).

(10) Nettesheim, P.H., D.G. Doherty, R.F. Newell and A. Hellman, J. Natl. Cancer Inst. 47, 1129 (1971).

(11) Oldiges, H., U. Glaser and D. Hochrainer, This workshop (1987).

Table 1: Incidence of primary tumors in the respiratory tract of male rats continuously exposed to Na-dichromate and a Cr VI/III oxide mixture by inhalation for 18 months and subsequent observation for a further year.

			$Na_2Cr_2O_7$			Cr_5O_{12}
		Controls	25 µg/m³	50 µg/m³	100 µg/m³	100 µg/m³
Rats per group	(n)	40	20	20	20	20
Rats survived 2 years	(n)	37	18	18	19	18
Rats with tumors	(n)	24	7	13	15	8
Total tumor rate	(n)	32	10	16	24	13
Lung tumors	(n)	-	-	-	3	1
Pharyns/Larynx	(n)	-	-	-	1	-

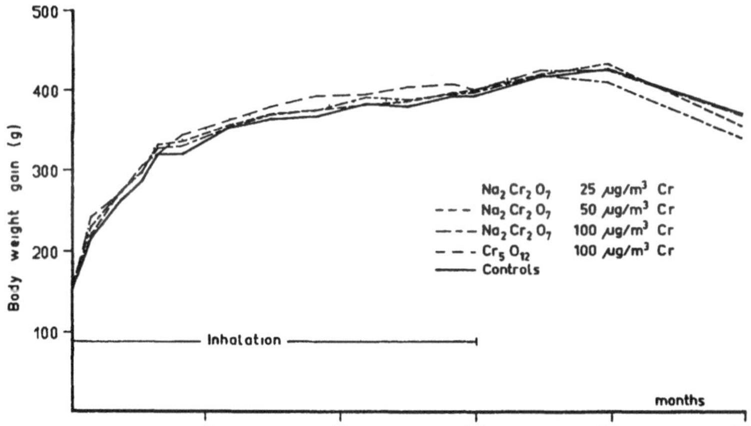

Fig. 1: Body weight gain (means) of the controls and the experimental groups continuously exposed to chromium VI aerosols ($Na_2Cr_2O_7$: sodium dichromate; Cr_5O_{12}: chromium oxide) for 18 months and subsequently housed conventionally for a further year.

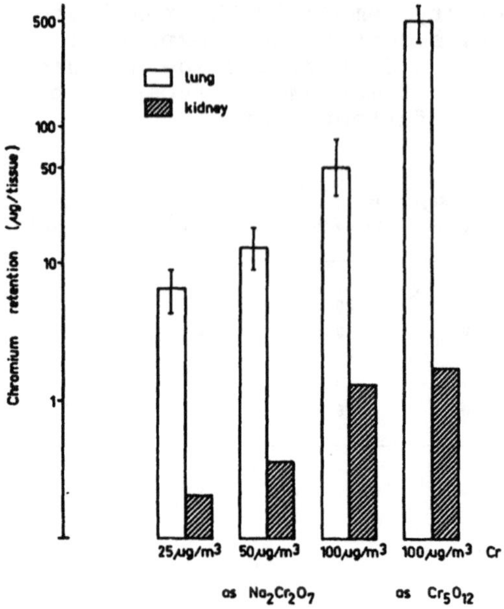

Fig. 2: Retention of chromium in lung and kidney (µg/tissue; X ± S.D. of 10 lungs or 6 kidneys), one year after termination of continuous exposures to $Na_2Cr_2O_7$ or Cr_5O_{12} aerosols for 18 months.

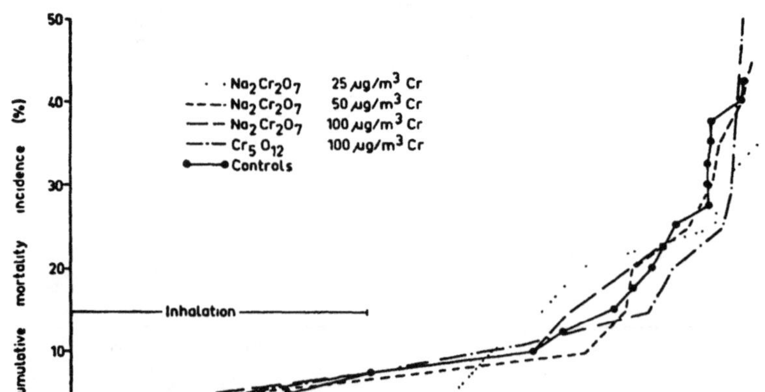

Fig. 3: Cumulative mortality incidence after 18 months continuous exposure to chromium VI aerosols ($Na_2Cr_2O_7$: sodium dichromate; Cr_5O_{12}: chromium oxide) in male Wistar rats.

LUNG TUMOR RISK ESTIMATES OF INHALED CADMIUM

G. Oberdörster
University of Rochester School of Medicine and Dentistry, Environmental Health Sciences Center, Dept. Biophysics, Rochester, NY 14642

INTRODUCTION

In the early 1960's it was reported that injected cadmium compounds can induce tumors at injection sites in rats and mice. It was later discovered that besides the site of injection tumors could also be induced in testes and pancreas. With regard to the respiratory tract, chronic inhalation of $CdCl_2$ aerosols were found to induce lung tumors in rats (1). Long-term inhalation studies with CdO, CdS and $CdSO_4$ designed after the study of Takenaka et al. resulted also in the induction of lung tumors (2). Thus, it appears that inhalation of many different cadmium compounds in the rat can induce lung tumors.

Epidemiological studies in cadmium exposed workers have recently been summarized (3). The latest study by Thun et al. (4) represents the best evidence that inhaled cadmium--in this case mostly CdO dust and fume--leads to an excess lung tumor risk. It was found that lung cancer mortality increased with increasing cumulative exposure. Although the authors corrected for smoking and simultaneous As exposure, the carcinogen assessment group of EPA felt that a possible exposure to As and cigarette smoking may have confounded the result (5). Consequently, they concluded that the evidence is limited that cadmium is a human lung carcinogen. Nevertheless, based on this study EPA estimated that the risk for lung cancer from lifetime exposure to 1 μg Cd/m^3 air is 1.87 x 10^{-3}. The objective of the following discussion is to use a dosimetric model of lung deposition and retention in rat and man to extrapolate lung tumor risk estimates from the rat study to humans.

METHOD

For the dosimetry of inhaled cadmium particles, both the initial deposition and the long term retention of cadmium in the respiratory tract have to be considered. Deposition and deposited dose rate in the respiratory tract depend on exposure concentration, breathing parameters (tidal volume, frequency), breathing mode (nose vs mouth) and airway geometry (length, diameter, branching angle, etc.). The accumulation of cadmium in cells of the respiratory tract resulting from its retention behavior depends on the biological halftime resulting from the metabolism of cadmium in the respiratory tract. Knowing these characteristics for rat and man and assuming first order clearance processes, one can then calculate the dose accumulating in the lung under constant exposure conditions. An equilibrium concentration A_E is reached when an exposure time of about 5 half-times has elapsed.

Knowledge about the deposition of inhaled particles in the respiratory tract is necessary for these calculations. Several investigators have performed studies in humans and rats to determine both total and regional deposition of different particle sizes in the respiratory tract. Based on such data, Yeh and Schum (6) and Schum and Yeh (7) have described a mathematical model to predict deposition of inhaled particles in rats and humans. Wojciak _et al_. (8) modified these lung model calculations and their model was _applied_ here to calculate deposition of cadmium containing particles in the human lung. Normal respiratory parameters compiled for a standard reference man (9) at rest were used.

Respective data for the rat were adapted to the conditions of the inhalation study by Takenaka _et al_. (1). The mean body weight over the exposure period of 18 _months was_ 400 g, and the empirical formulas by Stahl (10) were used for predicting minute breathing volume, respiratory rate and tidal volume. Cadmium containing particles in the ambient air are larger than those used in the rat study, they have an MMAD of 2.6 μm for rural areas and of 1.3 μm for industrial areas (11). For those particles, deposition efficiencies are different from the 0.55 μm particles for the rat study. The results of lung model predictions of deposition for a person living--at rest and during light activity--under these respective conditions in a rural area and in an industrial area show that pulmonary deposition is 10 percent (rural area) and 11 percent (industrial area) of the inhaled amount, respectively. By coincidence, for the two particle size distributions relative pulmonary deposition is practically identical under conditions of rest (10 and 11 percent). Because of this similarity and since the accumulation of pulmonary cadmium is of main interest for this extrapolation there is no need to consider rural areas and industrial areas separately under the aspect of particle size.

It has been shown, that the pulmonary retention of cadmium is quite different in rats and monkeys (12): The pulmonary retention halftime in rats is about 70 days, in monkeys it is about 780 days. Since humans are likely to retain cadmium more like monkeys than rats, a pulmonary retention halftime for cadmium of 780 days is assumed to exist in humans. Applying these specific pulmonary deposition and retention data for rat and man the accumulation of cadmium in the lung under a given exposure condition was calculated. Equilibrium concentration A_E in the lung tissue of the rat's lung is reached rather rapidly because of the short pulmonary retention halftime of only 70 days, whereas it takes much longer in the human lung with a retention half time of 780 days.

RESULTS AND DISCUSSION OF EXTRAPOLATION FROM THE RAT INHALATION STUDY

For extrapolation of the results of the rat inhalation study (1) the equilibrium concentrations in lung tissue occurring under the exposure concentrations of the study were calculated using the dosimetric approach described above. These equilibrium concentrations were plotted against the respective tumor incidences and the method of linear interpolation after Gaylor and Kodell (13) was adapted to calculate the upper limits of risk for low exposure concentrations of cadmium. Gaylor and Kodell showed that this approach leads to conservative estimates of risk at low doses comparable to those obtained from the multistage model or using safety factors.

To adapt the method of linear interpolation to the inhalation study the equilibrium concentration in the lung as obtained from the rat study was used. A logit analysis was applied to the data. Then an equivalent air concentration of cadmium for humans was calculated which--at a given particle size and breathing condition--will lead to this same equilibrium concentration in the human lung. This approach takes into account differences both in deposition and metabolism of cadmium in the lungs of rats and humans. An interpolation line according to (13) was calculated to obtain air concentrations of cadmium in the lower lung equilibrium concentration range representing the upper 95 percent tumor risk for these low levels.

Table 1 compares some results of risk estimates from this study with the lower risk estimate given by EPA. Since these are upper bound risk estimates only, inclusion of both the lower and upper bounds would show that estimates from both approaches overlap. Considering the completely different methods for obtaining these numbers the relatively good agreement of the results from both approaches seems to justify the use of this extrapolation modeling technique from the animal inhalation study. In addition, this method has the advantage of being adaptable to different environmental conditions--such as different particle sizes--and human activity conditions which can change the risk considerably. In general, the numbers show that lung cancer risk from inhaled environmental cadmium is very low.

Table 1 also gives the risk estimate for a one pack per day smoker based on the inhaled cadmium alone. The calculated risk of 1.1×10^{-2}, although significant, contributes still less than 9 percent to the total lung cancer deaths attributed to smoking, using estimates of lung cancer and cigarette consumption given in the Surgeon General's Report (14) and a recent NRC publication (15). Thus, measures to reduce smoking would certainly accomplish more in terms of decreasing lung cancer incidences than any preventive air pollution measure could do as has been pointed out repeatedly.

TABLE 1. Upper Bound Lung Tumor Risk from Lifetime Exposure to Environmental Airborne Cadmium

Population (Cd conc, ng/m^3)	Extrapolation from	
	Animal Study	Epidemiological Study
Rural (<1)	$< 3 \times 10^{-5}$	$< 3.5 \times 10^{-6}$
Urban (3.5)	1×10^{-4}	1.2×10^{-5}
Industrial (200)	6×10^{-3}	7×10^{-4}
1 pack/day smoker (500)	1.1×10^{-2}	1.75×10^{-3}

Estimated lung cancer deaths in 52.5×10^6 smokers (US-DHHS, 1982): 94,000

Calculated lung cancers due to Cd in 52.5×10^6 smokers (annually): 8,250

ACKNOWLEDGEMENTS

I am indebted to Dr. C. Cox for performing the logit analysis of lung tumor incidences. This paper is partly based on work supported by grants ES01247 and ES01248 from the National Institutes of Health.

REFERENCES

(1) Takenaka, S., H. Oldiges, H. Konig, D. Hochrainer and G. Oberdörster (1983) Carcinogenicity of cadmium chloride aerosols in Wistar rats. J. Natl. Cancer Inst., 70, 367-373.

(2) Oldiges, H., U. Glaser and D. Hochrainer (1987) Toxicity and carcinogenic risk after long term inhalation of cadmium compounds in Wistar rats. Proceedings of First European Meeting of Environmental Hygiene, Dusseldorf, May.

(3) Oberdörster, G. (1986a) Airborne cadmium and carcinogenesis of the respiratory tract. Scand. J. Work Environ. Health, 12, 523-537.

(4) Thun, M.T., T.M. Schnorr, A.B. Smith, W.E. Halperin and R.A. Lemen (1985) Mortality among a cohort of U.S. cadmium production workers - an update. J. Natl. Cancer Inst., 74, 325-333.

(5) EPA: Environmental Protection Agency (1985) Updated mutagenicity and carcinogenicity assessment of cadmium. Washington, DC (EPA-600/8-83-025F).

(6) Yeh, H.C. and M. Schum (1980) Models of human lung airways and their application to inhaled particle deposition. Bull. Math. Biol., 42, 461-480.

(7) Schum, M. and H.C. Yeh (1980) Theoretical evaluation of aerosol deposition in anatomical models of mammalian lung airways. Bull. Math. Biol., 42, 1-15.

(8) Wojciak, J.F., G. Oberdörster and R.H. Notter (1986) Dosimetric aspects of particle deposition in the respiratory tract of rat and man: Implications for extrapolating from rodent studies. In: Deposition and Clearance of Aerosols in the Human Respiratory Tract, ed. W. Hoffmann, Proceedings of Symposium, Salzburg, Austria, September (in press).

(9) ICRP: International Commission on Radiological Protection (1975) No. 23: Report of the Task Group on Reference Man. Pergamon Press, Oxford, New York, Toronto, Sydney, Braunschweig, pp. 151-173.

(10) Stahl, W.R. (1978) Scaling of respiratory variables in mammals. J. Appl. Physiol., 22, 453-460.

(11) Dorn, C.R., J.O. Pierce, P.E. Philips and G.R. Chase (1976) Airborne Pb, Cd, Zn, and Cu concentration by particle size near a Pb smelter. Atmosph. Environ., 10, 443-446.

(12) Oberdörster, G. and C. Cox (1986b) Lung retention of inhaled cadmium chloride aerosols in rodents and primates: A comparison. Toxicol. Letters Suppl., 31, 92.

(13) Gaylor, D.W. and R.L. Kodell (1980) Linear interpolating algorithm for low dose risk assessment of toxic substances. J. Environ. Path. Toxicol., 4, 305-312.

(14) US-DHHS: Department of Health and Human Services (1982) The Health Consequences of Smoking: Cancer. A report of the Surgeon General.

(15) NRC: National Research Council (1986) Environmental Tobacco Smoke. Measuring Exposures and Assessing Health Effects. National Academy Press, Washington, DC.

TOXICITY AND CARCINOGENIC RISK AFTER LONG-TERM INHALATION OF CD-COMPOUNDS IN WISTAR-RATS.

Oldiges, H., Glaser, U., Hochrainer, D.
Fraunhofer Institute for Environmental Chemistry and Ecotoxicology
D-5948 Schmallenberg-Grafschaft

INTRODUCTION

Earlier studies showed, that inhaled $CdCl_2$ induced primary lung tumors in Wistar rats (1.2). Also an increased mortality from respiratory cancer in workers exposed to cadmium has recently been reported (3). Furthermore an overview was given concerning airborne cadmium and carcinogenesis of the respiratory tract (4).

MATERIALS AND METHODS

This long-term inhalation study with male and female Wistar rats was designed to describe the toxicity and the carcinogenicity from 4 different Cd-compounds, namely Cd-chloride, Cd-oxide in the two forms, dust and fume, Cd-sulfate, Cd-sulfide and a combination of Cd-oxide/Zn-oxide. The planned exposure time towards the different Cd-aerosols was either 22 hours per day for 7 days per week and 18 months or in single cases for 40 hours a week and 6 months. The animals were exposed in horizontal flow inhalation chambers of 225 l volume. The aerosol-flow was 80 liters per minute. Each chamber contained two stainless steel wire mesh cages with 10 rats. The Cd-aerosols were produced by a reliable technique from earlier and new experiments (5). The $CdCl_2$, $CdSO_4$ and CdS aerosols were generated by jet nebulizers. The CdO dust aerosol was produced from a Cd-acetate solution and pyrolysis at 750 ^{o}C in a tube furnace and the Cd-fume aerosol by an electric arc generator. The particle size distributions were measured with a spiral centrifuge and the Cd-concentrations were determined by AAS (6, 7). The particle sizes ranged between 0.2-0.5 µm mass median aerodynamic diameter and the particles were thus respirable for rats.

RESULTS

In the literature no results were available on a long-term inhalation toxicity of Cd-compounds other than our own two experiments with $CdCl_2$ on rats. Because of lower water solubilities and expected lower toxicities we chose relatively high aerosol concentration for $CdSO_4$, CdS and CdO (8, 9). In some cases the chosen aerosol concentrations were too toxic. For instance the inhalation period with $CdSO_4$ at a concentration of 90 µg Cd/m^3 was finished for male rats after an exposure time of 13.5 months whereas the inhalation period for the female rats was 18 months. Thus, we found also a difference in the mortality between male and female animals (10). In case of CdS the aerosol concentrations of 270, 810 and 2430 µg Cd/m^3 had toxic effects, too. The inhalation period was stopped for 270 µg/m^3 Cd after 15.5 months, for 810 µg/m^3 after 6.8 months for male rats and after 9.7 months for female rats. For the highest concentration the exposure period was only about 3.5 months till the mortality rate was more than 25 %. Therefore we exposed 20 male and 20 female rats to the lower concentration of 270 µg Cd/m^3 for only 40 hours a week and 6 months.

CdO exists in two different forms as dust and fume. In our study we included both forms. The surface area of Cd-fume particles is greater than those of Cd-dust.

Therefore we assumed that Cd-fume may have a higher bioavailability and also a higher toxicity, but the present results do not support such assumptions.

Zinc is known to protect against the effects of Cd (11, 12). In two groups the rats were fed with a low-zinc diet.

In comparison to other groups exposed to the same Cd concentrations of 30 $\mu g/m^3$ but without zinc diet, the mortality appeared to be higher. The protective effect of zinc can be demonstrated from the exposure to combinations of Cd- and Zn-aerosols.

Remarkable results were found from the grossnecropsy. In all groups, not in the controls, we found a lot of lung nodules macroscopically; these were confirmed by histopathological examinations.

CONCLUSION AND SUMMARY

Based on the fact that this study includes the inhalation of water soluble and insoluble Cd-compounds, the toxicity must be considered under the aspect of bioavailability wich includes the concept of solubility. This concept also includes the importance of lung-clearance and phagocytic activity of the macrophages (13, 14). With regard to the carcinogenicity, we must conclude that all different compounds of Cd possess carcinogenic potencies.

In all Cd exposed groups the same kind of lung tumors were found namely adenomas, adencarcinomas, squamous cell carcinomas and mixed forms (Tab. 1). Therefore the carcinogenicity of inhaled Cd-aerosols must be related to the Cd ion.

ACKNOWLEGEMENTS

We thank Prof. Dr. U. Mohr, Medical Highschool, Hannover and Dr. Sh. Takenaka, Fraunhofer-Institute for Toxicology and Aerosolresearch, Hannover for the histopathological examination.

This study was supported by the German Bundesministerium für Forschung und Technologie, contract No. 01 VD, 183 and by the Eropean Commission, contract No. ENV-729-D (B).

Table 1: Preliminary histopathological results (Primary lung tumors)
(Date: 10.01.1987)

Sex	Cd-compounds	Nominal concentration ($\mu g\ Cd/m^3$)	Duration of exposure (months)	Bronchiolar/alveolar adenoma (n)	Bronchiolar/alveolar adenocarcinoma (n)	Squamous cell carcinoma (n)
♂♂	control	-	-	-	-	-
♀♀	control	-	-	-	-	-
♂♂	$CdCl_2$	90	6.0	2	1	1
♂♂	$CdSO_4$	90	13.5	1	4	2
♀♀	$CdSO_4$	90	18.0	6	4	4
♂♂	CdS	90	18.0	2	1	1
♀♀	CdS	90	18.0	-	1	1
♂♂	CdO-dust	90	7.1	1	-	-
♀♀	CdO-dust	30	18.0	1	1	1

REFERENCES

(01) Takenaka, S.,H.Oldiges, H.König, D.Hochrainer and G.Oberdörster (1983) Carcinogenicity of cadmium chloride aerosols in Wistar rats. J. Nat. Cancer Inst., 70, 373-376.

(02) Heering, H., G.Oberdörster, D.Hochrainer and H.P.Baumert (1979) Learning behavior and memory of chronically Cd exposed rats and organ distribution of Cd. Commission of the European Communities, Brussels, Doc. XII/ENV/64/97, 61-62.

(03) Thun, M.T., T.M.Schnorr, A.B.Simth, W.E.Halperin and R.A.Lemen (1985) Mortality among cohort of U.S. cadmium production workers An update. Journal National Cancer Institute, 74, 325-333.

(04) Oberdörster, G. (1986) Airborne cadmium and carcinogenesis of the respiratory tract. Scand. J. Work. Environ. Health 12, 523-537

(05) Hochrainer, D. and S.Soderholm (1986) A generator for cadmium oxide fumes aerosols. In: Aerosols Formation and Reactivity, Proceedings of the Second International Aerosol Conference, West-Berlin, 22-26 Sept. 1986, herausgegeben von der Gesellschaft für Aerosolforschung. Pergamon Press, Oxford, 1053-1056.

(06) Stöber, W. and H.Flachsbart (1969) Size-separating precipitation of aerosols in a spinning spiral duct. Environ. Sci. Tech. 3, 1280-1296.

(07) L'Vov, B.V. (1978) Electrothermal atomization - the way toward absolute methods of atomic absorption analysis. Spectrochim. Acta B 33, 153-193.

(08) Lauwerys, R.R. (1986) The toxicology of Cadmium. Commission of the European Communities, Brussels, EUR 7649 EN

(09) Löser, E. (1980) A two-year oral carcinogenicity study with cadmium on rats. Cancer Letters 9, 191-198.

(10) Kutzmann, R.S., R.T.Drew, R.N.Shiotsuka and B.Y.Cockrell (1986) Pulmonary changes resulting from subchronic exposure to cadmium chloride aerosol. J. Toxic. Environ. Health, 17, 175-189.

(11) Gunn, S.A., T.C.Gould and W.A.D.Anderson (1963) Cadmium-induced interstitial cell tumors in rats and mice and their prevention by zinc. J. Natl. Cancer Inst. 31, 745-759.

(12) Oberdörster, G., D.Hochrainer and H.Oldiges (1982) Prevention of CdO induced impaired lung chlearance by simultaneous ZnO inhalation. Toxicologist 2, 60.

(13) Oberdörster, G., H.P.Baumert, D.Hochrainer and W.Stöber (1979) The clearance of cadmium aerosols after inhalation exposure. Am Ind. Hyg. Assoc.H. 40, 443-450.

(14) Glaser, U., H.Klöppel and D.Hochrainer (1986) Biovailability indicators of inhaled cadmium compounds. Ecotox. and Environ. Safety, 11, 261-271.

POSTNATAL MOUSE HEART AND LUNG LYSOSOME RNAase LEVELS IN RESPONSE TO AIR POLLUTANTS

Orsi, E.V., G. Ames and M. Bavlsik
Biology Department, Seton Hall University, So. Orange, N.J. 07079,U.S.A.

INTRODUCTION

For some time we have been studying the response of cell culture systems to air samples from regions of strikingly different air quality. The high air pollution site (SHU) is located in So. Orange, New Jersey about 20 km from New York City. It is within 5 km of three major highways linking the New Jersey-New York metropolitan area with the Philadelphia region. There are at least 10 million inhabitants within a 50 km radius of SHU. It is at an elevation of about 30 meters and about 30 km from the Atlantic ocean. Some higher elevation is found 100-150 km to the northwest but it is an ineffective barrier to air trajectory pollution from the industrialized regions to the west[1,2]
The low air pollution site (BW-LAB) is in the Adirondack mountains of northern New York State about 50 km from the nearest express highway linking New York City with Montreal. It is in a valley of about 300 m elevation surrounded by peaks ranging from 1,000-1,600 m which tend to divert western pollution plumes[3]. It is in Wilmington,population of 800.
Since our results had shown detrimental effects on normal cells[4,5], we thought it useful to include animal systems for study. The use of lysosome response in cell pathology has been thoroughly documented since the early work of Allison on carcinogen uptake by lysosomes[6] and because our cell culture studies had indicated lysosome changes we decided to focus on lysosome enzyme changes in selected organs of mice born in the two regions.

MATERIALS AND METHODS

Mice Exposure to Ambient Air at Respective Sites

Timed pregnancy mice were purchased and upon arrival at SHU were divided randomly into two groups. One was taken to the mountains and the other driven in the same carrier box for the equivalent length of time (about 6 hours) and returned to SHU. In the first study, mice were kept in standard cages in the general laboratory area. In the second, cages were in an insulated chamber with constant 25°C and with 12 hours light periodicity. In both cases the same food, water and bedding were used. Five weeks after delivery, mothers and young were killed by neck dislocation, the carcass weighed, lung and heart removed and weighed followed by freezing and storage at -60°C prior to enzyme assay.

Lysosome Enzyme Levels of Hearts and Lungs

The marker enzyme was ribonuclease. In the first experiment total RNAse was measured. In the second, intra-lysosome and extra-lysosome activity was measured to monitor possible lysosome enzyme leakage.
The method is based on the release of soluble oligonucleotides following digestion of yeast RNA as adapted from various cell culture assay systems as follows[7,8] . Organs were homogenized and centrifuged for 15 minutes at 1475 x g. The supernatant was collected and centrifuged for 70 minutes at 20,200 x g. The sample was then usd or further processed for intra and extra-lysosomal samples. A 0.2 ml sample was added to 0.7 ml of sodium cacodylate (pH 6.0) and 1% yeast RNA (Sigma Chemicals, St. Louis, Mo.). After 30 minutes incubation, the reaction was stopped by adding 1.0 ml of $HClO_4$ containing 0.25% uranyl acetate. The solution was clarified by centrifugation and the supernatant measured at 260 nm.

Air Quality Measurements

The main parameter was the concentration of airborne particles inside, in the immediate vicinity of the cages and outside the laboratory. A portable small particle detector (Gardner Associates, Schenectady,N.Y.) was used. Counting was carried out with the chamber set for a minimum condensation nucleus size of 10^{-7} cm radius.

RESULTS

Air Quality at Rural and Urban Sites

The typical consistently higher particle counts obtained in the urban area is seen in Table 1. As expected there is a high correlation between outdoor and indoor levels. It is noteworthy that in the rural site indoor counts were often higher because of laboratory activity resembling that in the home[9]. In contrast, the urban laboratory levels were usually lower than outside particularly if air conditioning was in use. Although there are many other factors contributing to high particle levels such as local and distant industrial activity[2], vehicular traffic has been known as an important factor since the pioneer work of Schaefer[10]. This role is demonstrated both in Table 1 and 2. In Table 2, the fluctuation associated with holiday auto traffic is clearly seen along with the quick return to pre-holiday levels in the rural region.

TABLE 1

RANGE OF AVERAGE AIRBORNE PARTICLE COUNTS IN CN/CM^3*

	RURAL	URBAN
INDOOR	6,000-30,000	16,000-50,000
INDOOR**	600- 1,600	3,400- 8,000
OUTDOOR	900- 3,500	40,000-65,500
VEHICLES/HR	0-10	100- 1,000

*CONDENSATION NUCLEI

**AIR CONDITIONER IN USE

TABLE 2

HOLIDAY OBSERVANCE AND AIRBORNE PARTICLE LEVELS

	AVERAGE* RURAL	AVERAGE* URBAN
THANKSGIVING		
DAY BEFORE	1,000	10,000
HOLIDAY	7,000	30,000
DAY AFTER	1,500	11,000
CHRISTMAS		
DAY BEFORE	3,000	50,000
HOLIDAY	10,000	60,000
DAY AFTER	900	50,000
INDEPENDANCE DAY		
DAY BEFORE	700	28,000
HOLIDAY	1,400	--
DAY AFTER	600	66,000

*MULTIPLE COUNTS DURING DAY

Heart and Lung Lysosome Ribonuclease Activity

As seen in Table 3 mice born and raised in the rural area were heavier. In this first experiment there was also a suggestion that their hearts and lungs had more enzyme activity than those born and raised in the urban site. There was also an indication that the lungs of the mothers

in cleaner air had more ribonuclease activity.
In the second experiment (Table 4) the differences were even more pronounced. The rural young mice were again heavier but with hearts and lungs of lower weight but of higher RNA-ase activity. The mothers again showed a difference only in having lighter lungs with more RNA-ase levels if in the rural laboratory. The ratio of extralysome to total enzyme activity was essentially the same for all groups. Consequently lysosome leakage at a faster rate with urban animals was probably not involved.

TABLE 3

AIR QUALITY, CARCASS WEIGHT, HEART AND LUNG WEIGHTS AND RNA-ASE LEVELS

FIRST EXPERIMENT FALL	CFW MOUSE STRAIN URBAN	RURAL	P
NUMBER OF YOUNG	7	7	
AVERAGE WEIGHT (grams)	10.1	22.6	0.001
NUMBER OF MOTHERS	4	4	
AVERAGE WEIGHT (grams)	51.5	46.7	NS
AVERAGE HEART WEIGHT (grams) YOUNG	0.449	0.369	NS
AVERAGE HEART WEIGHT (grams) MOTHERS	0.726	0.596	NS
AVERAGE LUNG WEIGHT (grams) YOUNG	0.466	0.436	NS
AVERAGE LUNG WEIGHT (grams) MOTHERS	0.721	1.008	NS
HEART RNA-ASE (YOUNG) TOTAL	1.122	2.308	NS
HEART RNA-ASE (MOTHERS) TOTAL	0.975	2.384	NS
LUNG RNA-ASE (YOUNG) TOTAL	0.933	1.287	NS
LUNG RNA-ASE (MOTHERS) TOTAL	1.142	1.778	NS

URBAN = SO. ORANGE,NJ (NEW YORK CITY AREA)
RURAL = WILMINGTON NY (ADIRONDACK MTN.)
NS = NOT SIGNIFICANT AT P = 0.05
RNA-ASE = O.D. AT 260 nm/gm TISSUE

TABLE 4

SECOND EXPERIMENT SPRING	CF-1 MOUSE STRAIN URBAN	RURAL	P
NUMBER OF YOUNG	12	12	
AVERAGE WEIGHT (grams)	15.2	17.5	0.01
NUMBER OF MOTHERS	4	4	
AVERAGE WEIGHT (grams)	34.6	36.5	NS
AVERAGE HEART WEIGHT (grams) YOUNG	0.643	0.469	0.001
AVERAGE HEART WEIGHT (grams) MOTHERS	0.855	0.756	NS
AVERAGE LUNG WEIGHT (grams) YOUNG	0.682	0.419	0.001
AVERAGE LUNG WEIGHT (grams) MOTHERS	0.857	0.639	0.02
HEART RNA-ASE (YOUNG) TOTAL	1.088	1.622	0.001
HEART RNA-ASE (MOTHERS) TOTAL	1.128	1.280	NS
LUNG RNA-ASE (YOUNG) TOTAL	1.134	1.933	0.001
LUNG RNA-ASE (MOTHERS) TOTAL	1.015	1.670	0.001
EXTRA CELL/TOTAL HEART RNA-ASE			
YOUNG	52.9%	52.4%	
MOTHERS	51.3	48.2	
EXTRA CELL/TOTAL LUNG RNA-ASE			
YOUNG	54.2%	53.1%	
MOTHERS	52.4	48.6	

SUMMARY AND CONCLUSIONS

Mice born and raised in low air pollution areas were heavier but had hearts and lungs of lower wet weight. These organs however had more enzyme activity by weight. The lower activity in lysosomes of mice born and raised in a putative oncogenic environment is of particular interest because of the association of depressed lysosome enzyme activity and neoplasia[11,12]. Our results suggest that newborn mice are particularly sensitive to environmentally induced cell stress in contrast to adults of the same strain. Since we have recently found much higher levels of lead and copper in the urban site, measurement of lead and copper in mothers and progeny hearts and lungs are currently in progress[13].

REFERENCES

(1) Lioy,P.J.,G.T. Wolff and T.J. Kneip (1978) Toxic airborne elements in the New York metropolitan area, Air Poll. Con. Asso.,28,510-512.

(2) Rahn,K.A. and D.H. Lowenthal (1985) Pollution aerosol in the northeast:Northeastern-midwestern contributions, Science,228,275-284.

(3) Tester,J. (1964) Condensation nuclei counts on the slopes of Little Whiteface Mountain in New York state,Pure & Appl. Geophy.,57,181-188.

(4) Orsi,E.V.,P. Levins,G. Ames,L. Buonincontri,N. Holyk and R. Balciunas (1977) Human cell cultures as sentinel systems for monitoring air quality,637-650,In:Cell culture and its application,R.T. Acton and J.D. Lynn,Eds., Academic Press, New York.

(5) Orsi,E.V.,W.J. Zois and O.F. Baturay (1983) Human cell cultures response to simulated and natural levels of air pollution,In:In vitro testing of environmental agents:part B,A.Kolbar,T.K. Wong,L.D. Grant , R.S. DeWoskin and T.J. Hughes,Eds.,Plenum Publ. Co. New York

(6) Allison,A.C. and L. Mallucci (1964) Uptake of hydrocarbon carcinogens by lysosomes, Nature,203,1024-1027.

(7) Flanagan,J.F. (1966) Hydrolytic enzymes in KB cells infected with poliovirus and herpes simplex virus, J. Bacteriol.,91,789-797.

(8) Thacore,H. and D.A. Wolff (1968) Isolation and characterization of lysosome rich fractions from cultured cells, Exp. Cell Res.,29,266-276.

(9) Repace,J.L. and A.H. Lowrey (1980) Indoor air pollution, tobacco smoke and public health, Science,208,464-472.

(10) Schaefer,V.J. (1970) Auto exhaust, pollution and weather patterns, ASRC-SUNY pub. 121,31-33, reprinted from Bul. of the Atomic Scient.

(11) Poste,G.L. (1972) Sub-lethal autolysis, Exp. Cell Res.,67,11-16.

(12) Kelly,R.E. (1985) Sodium butyrate induced alterations in lysosomal enzyme activity, In Vitro Cellular & Dev. Biology,21,373-381.

(13) Orsi,E.V.,M. Bavlsik,R. Viera, M. Petersheim and O.F. Baturay (1986) Lysosome response and cytoskeleton alteration in cell cultures exposed to airborne lead, Third International Conference on Aerobiology,Basel, Experientia (in press).

III. BIOLOGICAL MONITORING

FORMIC-ACID EXCRETION IN URINE AS A BIOLOGICAL-MONITORING-PARAMETER IN AREAS WITH DIFFERENT AIR POLLUTION

Eikmann, Th.
Department of Hygiene and Occupational Health,
Technical University Aachen, Pauwelsstr.,D-5100 Aachen,FRG

INTRODUCTION

Within the framework of Biological Monitoring, the excretion of formic acid in urine indicates the degree of exposure of collectives to methanol, chlorinated methans and formaldehyde. The excretion of formic-acid in urine must be regarded as a relative problematic parameter of Biological Monitoring, because formic-acid is not only produced to a considerable extent by the break-down of substances with unstable methyl-groups and through the degredation of aminoacids but is also acquired to a considerable extent from food.
The aim of the study was therefore to prove possible influences on formic-acid excretion within general population, on the one hand, and to quantify the degree of exposure to the above mentioned substances in the general environment by means of this parameter, on the other.

MATERIALS AND METHODS

The evaluated data of the present inquiry were obtained from a larger epidemiological cross-sectional study, which was carried out from January to March 1982. The aim of this investigation was to find out possible effects of noxious substances in the air on the physical health of populations in polluted areal.

Adults of both sexes, from 60 to 65, were invited to participate in the investigation. A total of 1,274 volunteers took part in the study. Formic-acid was analysed by way of headspace analysis according to Angerer (1). In order to exclude possible deviations dependent on the time of day in the renal elimination of formic-acid, the latter was determined from 24-hour-urine. Samples, whose volume or creatinine-content were too low and indicated a non-exact collection of the 24-hour-urin, were excluded from the evaluation.

The districts of Leverkusen, Wiesdorf and Manfort and the complete township of Hürth were chosen as effected areals. The small town of Erkelenz in the lower Rhine-area served as the area of comparison. The area Leverkusen includes the second-largest formaldehyd production-plan in Western-Germany.

RESULTS

The frequency-distribution of separate values of formic acid shows an equally visible, distinct tendency to the left regarding both males and females. The 95-percentile-value of formic-acid excretion within the whole collective is about 75 mg per day, the 97,5 percentile-value at 100 mg per day.
If the whole collective is arranged to sex, one can observe a distinctly higher medium formic-acid excretion with males than with females. This difference can statistically secured as high significant.

If the whole collective is arranged according to sex and investigated areas the females of Leverkusen and Hürth show a higher formic-acid excretion than those of the female collective of Erkelenz. The medium excretion-values of the females of Leverkusen have a strong tendency to surpass those of the females of Hürth.
The male collectives of Leverkusen and Hürth also show a distinctly higher medium formic-acid excretion. These differences can however only be statistically secured through a comparison of the collectives of Leverkusen and Erkelenz.
The examination of possible influences of the investigated interfering parameters on the excretion of formic-acid, can only prove a dependency in connection with the smoking habits of the volunteers and the heating-systems of their homes. In parts this tendency is tendentious, in parts it can be statistically secured.

SUMMARY AND CONCLUSIONS

The medium formic-acid excretion of the collectives show a rather good agreement to some extent with the data of other investigators. A direct comparability is however not possible, due in part to the considerably different number of volunteers, to the different sample-material and to the different analytical methods. The distinctly higher medium formic-acid excretion of males compared to females is however striking. This difference has not been described so far and can be related in all probability to the usually higher work exposure and the higher mobility of males, as has shown before in earlier studies.
The differences in the medium formic-acid excretion values of the various collectives must be seen - in comparision with this - largely in accordance with the pollution concentration of the investigated areas.

REFERENCES

(1)Angerer, J. (1976) Gaschromatographische Bestimmung von Ameisensäure im Harn, in Form von Kohlenmonoxid. J. Clin. Chem. Clin. Biochem. 14, 73-77

EXPOSURE TO CADMIUM OF THE WEST-GERMAN POPULATION

U. Ewers, Medical Institute of Environmental Hygiene at the University of Düsseldorf, Gurlittstr. 53, D-4000 Düsseldorf 1 (FRG)

AVERAGE INTAKE AND ABSORPTION OF CADMIUM BY MAN IN THE FRG

Similar to the situation in other countries the diet and cigarette smoking represent the major sources of cadmium intake of the West-German population. Based on the median cadmium content of different food items and taking into account the average dietary habits of the West-German population the Federal Health Agency calculated the mean cadmium intake of males to be 0.2425 mg/week and that of females to be 0.1882 mg/week (1). The cadmium concentration in the tobacco of German commercial cigarettes has been studied by Szadkowski et al. (2), Müller (3) and Scherer & Barkemeyer (4). According to these studies the cadmium concentration in cigarette tobacco varies between 0.2 and 2.0 ppm or 0.5 - 1.5 µg Cd/cigarette. The average cadmium content in the mainstream smoke of German filter cigarettes is about 0.1 µg Cd/cigarette or less. In plain cigarettes the cadmium content of the mainstream smoke usually exceeds 0.1 µg/cigarette. 80 - 95 % of the mainstream smoke cadmium occur in particulate form. Gas-phase cadmium contributes between 5 and 20 % to the total amount of cadmium in the mainstream smoke.

Assuming a gastrointestinal absorption of 5 % and a pulmonary retention of 30 % the average cadmium intake and absorption of the West-German citizen can be characterized as shown in Table 1. It is evident that the diet represents the major source of cadmium intake and absorption in non-smokers, whereas in cigarette smokers the inhalation of cigarette smoke may contribute about 50 % of the total cadmium amount absorbed resulting in a nearly twofold increased cadmium absorbtion compared to non-smokers.

Taking into account a mean cadmium concentration of 1 µg/l in drinking water, the cadmium intake from water and water-based beverages is likly to average 1 - 2 µg/day. The concentrations of airborne cadmium vary between 1 and 3 ng/m^3 in rural and urban areas (5). Breathing 15 m^3 of air per day would thus result in an uptake of 0.02 - 0.05 µg/day. Assuming a deposition of 30 % in the lung tissue (the remainder being exhaled) and a complete absorption of the deposit results in a pulmonary absorption of less than 0.01 up to 0.02 µg/day. Absorption of cadmium in normal respiration is, therefore, small compared with the amount absorbed from food or from inhalation of cigarette smoke.

The FAO/WHO Joint Expert Committee on Food Additives recommended a provisional maximum tolerable weekly intake for cadmium from all sources of 400 - 500 µg (6). The data presented in Table 1 show that the average cadmium intake by the general population in the FRG is well below this limit. It should be noted, however, that the amount of cadmium ingested by individuals is likely to vary widely with individual habits. People consuming, for example, large quantities of certain species of shellfish (particularly crab brown meat and pastes), or eating some types of vegetables grown on soil contaminated by cadmium fall-out near lead and zinc smelters or repeatedly fertilised with sewage sludge containing cadmium, are likely to have intakes substantially in excess of the average and perhaps in excess of the FAO/WHO recommended level.

Table 1. Average cadmium intake and absorption in the general population of West-Germany

Source	Cadmium intake (µg/Tag)		Cadmium absorption (µg/Tag)
Diet and drinking water [a]	34.6 (M) 26.9 (F)	GIT / Absorption rate 5 %	1.5
Air [b]	0.04	Lung / Deposition rate 30 % complete absorption	0.02
cigarette-smoke (20 cig./day) [c]	2.0	Lung / Deposition rate 50 % complete absorption	1.0

a) Data according to ref. (1).
b) Taking into account air cadmium levels of 2.5 ng/m³ and a breathing volume of 15 m³ per day.
c) Assuming an inhalation dose of 0.1 µg per cigarette.

LEVELS OF CADMIUM IN HUMAN BIOLOGICAL MEDIA AS INDICATORS OF EXPOSURE

In the last years, several studies were conducted in the FRG to determine the levels of cadmium in blood, urine, and hair in the occupationally non-exposed general population. The more recent studies applying modern analytical techniques and careful quality control have been reviewed by Ewers et al. (7). Additional data concerning cadmium levels in blood, urine, and hair of children have been presented by Brockhaus et al. (8) and Wilhelm et al. (9). From these studies the normal ranges of the cadmium concentrations in the above named biological media can be derived. The upper limits of the normal ranges (95. percentile values) are summarized in Table 2. It should be noted that the mean values usually are substantially below these "upper normal limits".

The available data show that the cadmium concentrations in blood, urine, and hair are influenced by a number of factors:

- Cigarette smokers have significantly higher blood cadmium levels than non-smokers;
- Adult non-smokers have higher blood cadmium levels than children;
- Adults have higher urine cadmium levels than children;
- People living near lead and zinc smelters have, on average, higher levels of cadmium in blood, urine and hair than otherwise comparable population groups;
- Women and young girls tend to have higher urine cadmium levels than men and boys, respectively, of similar age.

Data on the cadmium concentrations found in placental tissue of mothers living in different areas of the FRG have been presented by Thieme et al. (10) and Krause et al. (11). The mean values vary between 40 and 60 ng/g

Table 2. Normal ranges of the cadmium concentrations in human biological media as indicators of exposure.

Medium	Normal range (95. percentile)	Population group
Blood	0.5 µg/g	children
	1.0 µg/g	adult non-smokers
	5.0 µg/g	adult smokers
Urine	1.0 µg/g creatinine	children
	2.0 µg/g creatinine	adults
Head hair	0.5 µg/g	children

dry weight with 95. percentiles being in the range of 100 - 150 ng/g. The mean values reported by Thürauf et al. (12) are much too high, presumably due to a systematic analytical error. Data on the cadmium levels in human liver tissue have been reported by Thürauf et al. (13). The median values range from 1.5 to 2.6 µg/g wet weight.

LEVELS OF CADMIUM IN HUMAN KIDNEY CORTEX AS INDICATOR OF CADMIUM CONCENTRATION IN THE CRITICAL ORGAN

The critical organ after chronic cadmium exposure is the kidney. Data on the distribution of cadmium in kidney cortex (CdKc) can be used to assess whether the critical cadmium concentration in renal cortex, upon which tubular proteinuria is initiated, is reached in a fraction of the general population. Estimates of the critical CdKc concentration range from 160 to about 400 µg/g (wet weight) with 200 µg/g being the most widely accepted value (13,14). It should be noted, however, that this does not imply that 200 µg/g would give rise to renal dysfunction in <u>all</u> exposed persons. As frequently can be observed in biological systems, at a certain low concentration or dose only a fraction of the exposed individuals will show signs of effects. Therefore, Kjellström et al. (14) suggested to use the term "population critical concentration" (PCC) instead of "critical concentration (on an individual basis) when assessing population based data. The PCC should be determined in relation to a particular response rate in the population. According to Kjellström et al. (14), the PCC-10 (i.e. the CdKc concentration that would give rise to tubular proteinuria in a fraction of 10 % of an exposed population) is in the range of 180 - 220 µg/g.

In total, five autopsy studies have been performed to evaluate the distribution of CdKc levels in residents of the FRG. The results of these studies are summarized in Table 3. The median CdKc values consistently were found to be about 20 µg/g. Individual values range up to about 100 µg/g. Drasch (16) reported on several cases with CdKc levels above 100 µg/g, two of these cases being in the range of the PCC-10. The studies of Drasch (16) and Hahn et al. (17) show a marked increase of CdKc up to about age 50 years and a decrease of CdKc at higher ages. All studies, in which the smoking histories of the autopsied individuals were recorded, consistently show significantly higher CdKc levels in cigarette smokers compared to non-smokers. The increase of CdKc is dose-

Table 3. Cadmium concentrations in kidney cortex (CdKc). Results from autopsy studies conducted in different areas of West-Germany.

Area	N	CdKc (µg/g wet weight) Median	Range	Reference
Erlangen/ Nürnberg	89	21.0	2 - 115	Thürauf et al. (19)
München	263	22.0	max. 230	Drasch (15)
München	47	25.7 [b)]	1 - 99	Summer & Heilmair (20)
Goslar	28	19.3	3 - 125	Thürauf et al., (13)
Erlangen	50	12.2	1 - 58	
Düsseldorf/ Duisburg	388	17.1	2 - 95	Hahn et al. (16)

[a)] arithmetic mean

dependent on a group basis. People from cadmium-polluted areas have, on average, higher CdKc levels than people from non-polluted areas (13,16). According to a study by Drasch (17), there has been a substantial increase of the average cadmium concentration in human kidney cortex since the beginning of this century.

CONCLUSIONS

The average cadmium intake and the levels of cadmium in human biological media of inhabitants of the FRG are well comparable with those found in other industrialized countries, e.g. USA, Sweden, Italy and The Netherlands. Higher cadmium levels in blood, urine, and kidney cortex have been found in Japan and in some areas of Belgium. From the distribution of the urine cadmium concentrations in the adult population of the FRG it can be concluded that more than 98 % are below the maximal individual biological limit of 5 µg/g creatinine recommended by WHO (18). In occupationally non-exposed individuals the median cadmium levels in kidney cortex are in the range of 20 µg/g wet weight. The 95. percentile is about 50 µg/g. Levels above 100 µg/g or ranging up to the PCC-10 are scarcely found in occupationally non-exposed individuals.

REFERENCES

(1) ZEBS (= Zentrale Erfassungs- und Bewertungsstelle für Umweltchemikalien des Bundesgesundheitsamtes) (1984) Arsen, Blei und Cadmium in und auf Lebensmitteln. ZEBS Hefte 1/1984.

(2) Szadkowski, D., et al. (1969) Zur ökologischen Bedeutung des Schwermetallgehaltes von Zigaretten. Arch.Hyg.Bakteriol. 153, 1-8.

(3) Müller, G. (1979) Schwermetallgehalte (Cd, Zn, Pb, Cu, Cr) im Tabak häufig in der BR Deutschland gerauchter Zigaretten. Chem.Ztg. 103, 133-137.

(4) Scherer, G., and H. Barkemeyer (1983) Cadmium concentrations in tobacco and tobacco smoke. Ecotox.Environ.Safety 7, 71-78.

(5) Buck, M. et al. (1982) Die Entwicklung der Immissionsbelastung in den letzten 15 Jahren in der Rhein-Ruhr-Region. Landesanstalt für Immissionschutz des Landes Nordrhein-Westfalen (LIS), LIS-Bericht Nr. 18, Essen.

(6) WHO (1972) Technical Report Series 505. Geneva.

(7) Ewers, U., et al. (1988) Exposure to cadmium of the West-German population - Results of biological monitoring studies 1980 - 1986. In: M. Stoeppler and M. Piscator (Eds.), Environmental Toxin Series, Vol. 2 - Cadmium. Springer Verlag, Berlin Heidelberg New York (in press).

(8) Brockhaus, A., et al. (1988) Exposure to lead and cadmium of children living in different areas of North-West Germany: results of biological monitoring studies 1982 - 1986. Int.Arch.Occup.Environ. Health (in press).

(9) Wilhelm, M., et al. (1987) Variables influencing cadmium concentrations in hair of pre-school children living in different areas of the FRG. Int.Arch.Occup.Environ.Health (in press).

(10) Thieme, R., et al. (1986), 10-Jahres Vergleich an Ablagerungen von Schwermetallen in der menschlichen Pazenta. Geburtsh. u. Frauenheilk. 46, 180-183.

(11) Krause, Ch., et al. (1987) Blutcadmiumgehalte bei Mutter und Kind. Schr.-Reihe Verein WaBoLu 74, 165-203. G.Fischer Verlag, Stuttgart.

(12) Thürauf, J., et al. (1975) Der Cadmium-Gehalt der menschlichen Plazenta. Int.Arch.Occup.Environ.Health 36, 19-27.

(13) Thürauf, J., et al. (1986) Cadmium concentrations in autopsy material from differently polluted areas of West-Germany (FRG). Zbl.Bakt.Hyg. B 182, 337-347.

(14) Kjellström, T., et al. (1984) Conceptual problems in establishing the critical concentration of cadmium in human kidney cortex. Environ.Res. 33, 284-295.

(15) Drasch G (1982) Kadmium-Belastung im südbayerischen Raum. Münch. med.Wschr. 124:1129-1132.

(16) Hahn, R, et al. (1987) Cadmium in kidney cortex of inhabitants of North-West Germany: its relationship to age, sex, smoking and environmental pollution by cadmium. Int.Arch.Occup.Environ.Health 59:165-176.

(17) Drasch, G. (1983) An increase of cadmium body burden for this century - an investigation on human tissues. Sci.Total Environ. 26, 111-119.

(18) WHO (1980) Technical Report Series 647, Geneva.

(19) Thürauf,J., et al. (1981) Zur gegenwärtigen Belastung der Bevölkerung mit Kadmium. Fortschr.Med. 99:1312-1317.

(20) Summer, K.H., und H. Heilmaier (1985) Metallothionein und Kadmium in der menschlichen Nierenrinde: Einfluß des Rauchens. In: Wege und Wirkungen von Umweltchemikalien. Tagung der Arbeitsgemeinschaft der Großforschungseinrichtungen (AGF) am 12. und 13. Dezember 1985 im Wissenschaftszentrum, Bonn - Bad Godesberg. Manuskript, Eigenverlag.

EVALUATION OF ALVEOLAR BURDEN OF MINERAL FIBRES IN CASES OF OCCUPATIONAL AND NON OCCUPATIONAL EXPOSURE TO ASBESTOS

Friedrichs,K.H.*, G. Chiappino**, A. Forni** and A. Rivolta**

* Medical Institute of Environmental Hygiene at the University of Düsseldorf, FRG

**Research Centre on Biological Effects of Inhaled Dust Institute of Occupational Health, University of Milan, Italy

INTRODUCTION

Counts of fibres from lung tissue are of increasing importance in medicolegal cases in Germany. For this purpose, different fibrerecovery techniques have been developed and different microscopic methods and instruments have been employed. Besides the examination of dust samples from lung tissue few data have been presented on the fibre content in broncho alveolar lavage (BAL) samples. This study was designed to determine the fibre concentration, size distribution and composition of fibres in BAL-samples from individuals with different occupational exposure.

MATERIALS AND METHODS

Samples were kept under formalin until examination. Prior to filtration, the organic material was removed with KOH, the sediment was then treated in an ultrasonic bath for a short period. The NP-Filters were evaluated with a SEM at a magnification of 6000x by counting, measuring and analyzing fibres directly on the screen, and a TEM at a normal magnification of 16000x. Both microscopes are fitted with energy dispersive x-ray spectrometers. Additionally, the TEM has the capabilities for SAED. In these cases the magnification was raised to 31000 x. 250 fields per sample were evaluated with the SEM. Ten grid openings per grid were analyzed with the TEM for "particles with nearly parallel long edges, which had an aspect ratio of 3 : 1 and more".

In total, 57 BAL-samples, which were classified into the following groups were examined:

- 19 cases of current and 18 cases of ceased occupational asbestos exposure
- 5 cases of occupational exposure to MMMF
- 9 cases of occupational exposure to other industrial dusts
- 6 cases of no known occupational exposure

RESULTS

In comparison to the less exposed groups 3, 4 and 5 the total fibre concentration is higher in the first two classes. The standard deviation is large in the classes 1 and 2. The mean fibre concentration is also higher for TEM-data by a factor of 2.4 - 4.3 compared with the SEM. The linear relation between the microscopes with a correlation coefficient of 0.84 seems to be rather strong. The limitations of SEM, however, become evident when more substantial information on the fibres is considered. Since fibres with diameters below 0.2 um cannot be analyzed properly with the SEM, the following results are presented only for the TEM examinations.

TEM ANALYSIS OF BAL SAMPLES

GROUP	FIBRES /ml SUSPENSION				
	CHRYSOTILE	AMPHIBOLES	ASBESTOS-BODIES	OTHER	ASBESTOS /OTHER
CURRENTLY ASBESTOS EXPOSED (n = 19)	20780	19190	1750	11050	3.3
PREVIOUSLY ASBESTOS EXPOSED (n = 18)	19740	11940	1030	10150	3.3
EXPOSURE TO MMMF (n = 5)	4130	6780	—	5020	2
EXPOSURE TO OTHER INDUSTRIAL DUSTS (n = 9)	12730	5190	—	8540	2
NO OCCUPATIONAL EXPOSURE (n = 6)	4910	2120	—	5350	1.3

Figure 1 : Mean fibre concentration per ml BAL suspension in five groups with different occupational exposure analyzed with the TEM

A more detailed survey over the different fibre types is shown in Figure 1 for the TEM-data. The longest "fibres" are asbestos bodies with diameters that correspond to those of the amphibole group. The thinnest fibres are chrysotile, which show a decrease of their average length from class 1 to class 5. It is notable that "other" fibres are similar in length and in diameter in nearly all the classes.

It has been tried to gain some informations on the biological relevance of the results by grouping the fibre dimensions according to existing models of carcinogenicity, for example that established by Stanton et al (1).

LENGTH

DIAMETER	< 1	1 - 4	4 - 8 *	8 - 64 *
- 0.1	42 53	11 14		
0.11-0.25	19 15	24 18	1	
0.25-0.5		3		
0.5-1.5				

OTHERS * STANTON - RANGE

Figure 2 : Concentration of "other" fibres (in %) grouped in different length- and diameter classes for asbestos exposed and non exposed humans
Size classes grouped after Stanton

From Figure 2 it can be seen that the fibre content per size class of those, for this purpose combined groups 1 and 2 as the asbestos exposed, and the groups 3,4 and 5 as the non-exposed on the other side, is quite similar for "other" fibres. Furthermore, the figures are not placed in the "dangerous" range. Figures 3 and 4 indicate that in the asbestos exposed group the fibre content (%) in the Stanton range is higher compared with the reference group.

LENGTH

DIAMETER	< 1	1 - 4	4 - 8 *	8 - 64 *
- 0.1	25 31	32 39	8 2	8 3
0.11-0.25	3 4	12 15	5 3	2 1
0.25-0.5		1 1	1	1
0.5-1.5			1 1	1

CHRYSOTILE   * STANTON - RANGE

Figure 3 : Chrysotile concentration (in %) of asbestos exposed and non exposed groups for different length and diameter classes
Size classes grouped after Stanton

LENGTH

DIAMETER	< 1	1 - 4	4 - 8 *	8 - 64 *
- 0.1	11 22	20 27	5 2	1
0.11-0.25	9 4	26 34	7	5 2
0.25-0.5		6 7	5 2	3
0.5-1.5			1	1

AMPHIBOLES   * STANTON - RANGE

Figure 4 : Amphibol fibre concentration (in %) of asbestos exposed and non exposed groups for different size classes
Size classes grouped after Stanton

CONCLUSIONS

Based upon the preliminary results of the present study it may be concluded that the differentiation between occupationally exposed and non exposed individuals should be possible by analyzing asbestos fibres, presumably the longer ones. These examinations can only be performed by TEM-analysis since fibres of all diameters seem to be of equal importance.

REFERENCE

(1) Stanton,M.F.and M.Layard (1978) The Carcinogenicity of fibrous minerals
In: Proceedings of workshop on asbestos: Definitions and measurement methods.
Gaithersburg,Md.: Natl Bur Std, 143-151

CHANGES OF BIOCHEMICAL PARAMETERTS FOLLOWING SHORT-TERM EXPOSURE TO AIRCRAFT-NOISE.

Marth, E., *E.Gallasch, **G.F.Fueger and J.R.Möse
Institute of Hygiene, *Institute of Physiology and ** Dep.of Nucl.Medicine of the University of Graz, Univpl.4, A-8010 Graz, Austria

INTRODUCTION

Examination of the physiology of human sound processing showed that the impulses triggered by sound stimuli reach the auricular cortex via auditory nerve. At the same time, sound stimuli branch off and reach the reticular formation. Reticular formation, hypothalamus and pituitary are in close relationship with each other and together adjust and coordinate the autonomic nervous and endocrine systems(6). The effect of different stressors on adjustment and coordination is determined by different parameters. The most important factors are briefly mentioned here:

1. Release of adrenocorticotropic hormone(ACTH) and cortisol, as well as catecholamines (2);
2. Change in the heart rate, blood pressure and respiratory rate as direct consequence of an increase in catecholamines;
3. Reduction of blood supply to the skin and consequent lowering of skin temperature;
4. Increase of metabolism characterized by an increased demand for energy (1).

If sound affects an organism without prior warning, an alarm reaction is triggered. At that point, all safeguards necessary for the protection and preservation of life are mobilized. This sudden physical stress is transmitted to the anterior lobe of the pituitary in the form of humoral or nervous stimuli. Then ACTH is realeased.
The regulatory elements of the hypothalamic-pituitary-adrenal cortical system for regulation of cortisol secretion, a closed loop negative feedback control system with open loop transients. Secretion of ACTH is stimulated by CRF (corticotropin releasing factor) and inhibited at the pituitary level by cortisol. Cortisol also acts on the hypothalamus or other parts of the brain to inhibit secretion of CRF (4).
The influence of noise on various biochemical parameters was studied in animal and clinical-chemical laboratory experiments. The duration of noise exposure in the experiments was several hours, in some cases months.
Special attention was given to those parameters which are directly influenced by the effect of ACTH, i.e. blood sugar and lipid metabolism. Blood sugar as well as free fatty acids produced by the hydrolisis of the triglycerides, are oxidized by β-oxidation or by the citric acid cycle. In the course of the oxidative phosphorylation, energy is produced in the form of ATP, which is decisive significance in fight or flight reactions.

MATERIALS AND METHODS

Experimental arrangement

The present study investigates the effects of aircraft-noise, how even short-term exposure to noise of 105 dB(A) can provoke ACTH-release in individuals and which biochemical parameters are altered. 14 female and 11 male subjects ranging in age between 21 and 43 yeares, with an mean age of 25 years were used in the experiments. An examination prior to the experiment showed that all of them were healthy with regard to metabolism. At the beginning of the test, an intravenous line was inserted in order to prevent distortion of the results as a consequence of a venous puncture.
Groups of 3 subjects each were formed. These groups were individually exposed to the aircraft-noise simulator (Fig.1). The simulator produced the noise level found in field studies of low-level flights reached a maximum of 105 dB(A). The exposure to this maximum level was limited to approximately 3 seconds. Prior to and 5 minutes following the noise exposure, blood samples for an analysis of the biochemical parameters were drawn from the venous line of the test persons.

Adrenocorticotropic hormone

ACTH was analysed by means of a radio-immuno assay. The principle of this method is based on the competition for free binding sites of the specific ACTH-antibody occurring between labeled ACTH and unmarked ACTH sample (5).

RESULTS

The analysis of ACTH prior to the noise exposure showed a normal concentration in 96% of all cases, while initial values were slighly pathological in 4%. A normal value is generally defined as not exceeding 70 pg/ml, according the circadian rhythms. The mean value was 25pg/ml. Following the exposure to aircraft noise, the concentration increased by 100 % when compared to the initial value (Fig.2). In 28% of the cases, pathological concentrations were found. In 48% of the cases, ACTH concentration more than doubled when compared to the initial values. The mean value was 46pg/ml and the standard deviation 39pg/ml (3). The comparison of the two mean values showed a statistical significant difference.
In comparing triglyceride concentrations prior to and following noise exposure it was observed that the concentration decreased in 85% of all cases. This was not expressed in a significant difference in mean values because the time period from exposure until the drawing of the sample was too short. However, if one uses trend tests in order to carry out trend analyses, a highly significant difference between the values prior to and following the exposure can be calculated.
Cholesterol concentration increased in 86% of all cases. While the comparison of mean values did not result in a significant statistical difference, the trend analysis again showed a significant tendency toward an increasing concentration.
Glucose levels were variable. The blood sugar values reacted to noise exposure in two phases. Following exposure, a temporary increase was noted, followed by a gradual decrease. The diagram shows

blood sugar levels under normal conditions and under stress (Fig.3). Under normal conditions, glucose released by the activity of glucagon is mostly metabolized by the brain, a small part by muscle and adipose tissue. Under stress, the following process was observed: first, more liver-glycogen is broken down as a consequence of an increased glucagon activity, resulting in increased glucose levels in the plasma. In the second phase, an increased glucose consumption was noticed, expressed by a reduced blood sugar level.

SUMMARY AND CONCLUSIONS

In low-level training flights, military fighter planes produce a high-intensity noise level without prior warning. This sudden psychological stress results in a prompt reaction in the form of ACTH-release. The level of the ACTH-concentration varies according to the individual. Each stress situation results in reactions designed to aid in the preservation of life. Most importantly, energy is provided in order to enable the system to cope with intensified metaboilsm. By oxidation of glucose and free fatty acids, the organism forms energy in form of ATP relatively rapidly.

The free fatty acids are formed by the hydrolytic breakdown of triglycerides. Our experimental model also showed a decrease in the concentration of triglycerides in the serum.

Glucose is mobilized rapidly so that a temporary increase of the blood sugar level was found. Later the blood sugar level decreases significantly as a result of the increased demand in the various organs.

Short-term high-intensity aircraft noise resulted in changes in the biochemical parameters through an influence on the endocrine system. ACTH was the hormone at the beginning of a chain reaction whose concentration indicated whether the intensity of the flight noise was able to trigger the chain reaction characterizing a stress situation.

REFERENCES

(1) Jansen,G.(1961) Effects of noise on the vegetative nervous system of man, Ger.Med.Mon., 12-13.

(2) Krypter, K.D.(1970) The effects of noise on man, Academ.Press New York, London.

(3) Marth, E., E.Gallasch, F.G.Fueger and J.R.Möse(1987)Fluglärm: Veränderung biochemischer Parameter, Zbl.Bakt.Hyg.,I.Abt.Orig.B in press

(4) Martin, J.B. and S.Reichlin(1977) Clinical Neuroendocrinology, ed. Davis, F.A., Philadelphia

(5) Orth, D.N. and W.E. Nicholson(1973) Biological and immunologic characterization and physical separation of ACTH and ACTH-fragments in the ectopic ACTH-syndrome, J.Clin.Invest.,52,1756

(6) Scheidt,R., R.Bartsch, and Chr.Brückner(1986) Die extraaurale Wirkung von Lärm- Eine Literaturübersicht, Lärmbekämpf. 33, 151-159

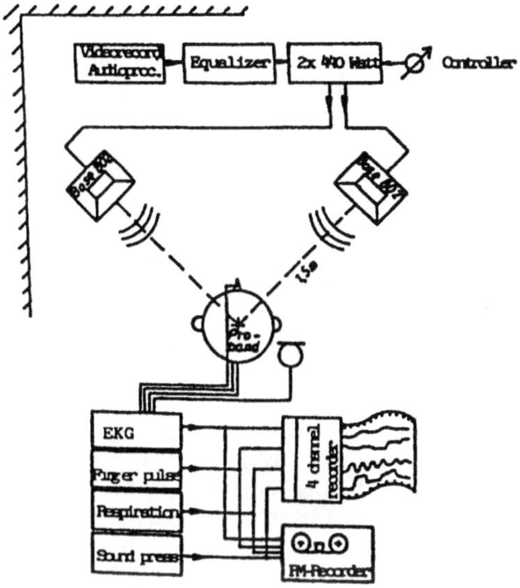

Fig. 1

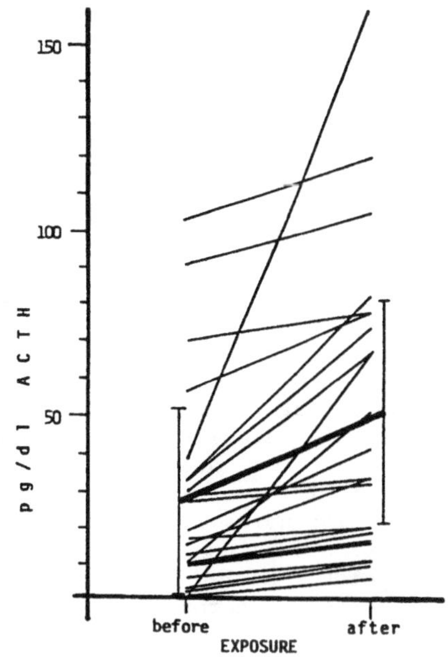

Fig. 2

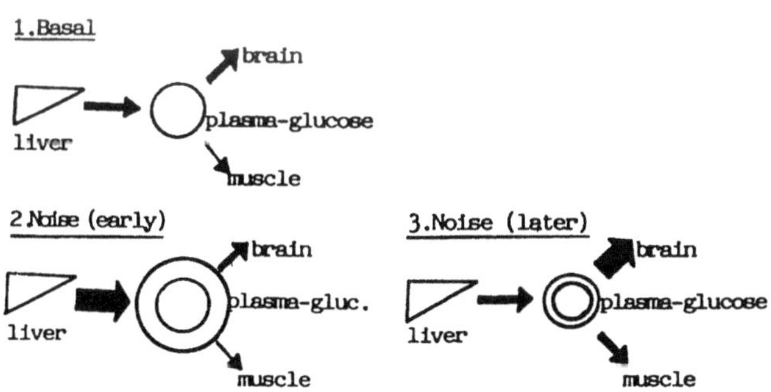

Fig. 3

DEPOSITION AND SHORT-TERM CLEARANCE OF AEROSOL PARTICLES IN THE HUMAN RESPIRATORY TRACT

W. Stahlhofen
Gesellschaft für Strahlen- und Umweltforschung mbH., München, Institut für Biophysikalische Strahlenforschung, Paul-Ehrlich-Str. 20,
6000 Frankfurt am Main, FRG

INTRODUCTION

Health effects resulting from particle inhalation are related to the sites of deposition in the respiratory tract and the pattern of clearance from each site. When particles are inhaled, two distinct phases of clearance from the thorax are usually observed. It is generally assumed that the fast phase, which is complete within about a day, represents mucociliary clearance of particles deposited in the tracheobronchial tree, and the slow phase represents clearance of the alveolar deposit. This assumption forms the basis of measurements of regional deposition using radioactively labelled aerosols. To investigate these assumption, lung clearance experiments are being conducted in which particles labelled with gamma emitting radionuclides are administered as "bolus" of aerosol, injected at a predetermined point in the breathing cycle, to restrict particle deposition to a relatively small region of the lung.

MATERIAL AND METHODS

With an inhalation technique described in a previous paper it is possible to inject aerosol boluses with defined volumes at any moment during inhalation into the inspired air (1).
For technical reasons the method was optimized using pulses of about 50 cm^3 volume. By varying the point of injection during inspiration and by holding the breath at end-inhalation deposition of particles in different volumetric lung depths was achieved. Other parameters which have been controlled were the particle size and the respiratory flow rate. By this way it was expected that particle clearance can be measured as function of the site of particle deposition in the lung. Monodisperse spherical iron oxide particles between 1.2 to 3.5 µm aerodynamic diameter were produced by atomization of an aqueous colloidal solution of Fe_2O_3 with a spinning top generator. For radioactively labelling the Fe_2O_3-colloid was mixed with colloidal ^{198}Au.

Four male nonsmokers with normal lung functions volunteered in the experiments. Starting from ERV, variable volumes of more than 900 cm^3 of clean air were inhaled at a constant flow rate of 250 cm^3. At this level a 50 cm^3-bolus containing radioactively labelled particles was injected by means of a valve system and inspiration of clean air was continued until a fixed tidal volume of 1000 cm^3 was completed.
Some experimental parameters used in the aerosol pulse technique are defined in accordance with Fig. 1 as follows: The front depth of a pulse is the volumetric distance between larynx (l) and the front of the pulse at end-inspiration under the assumption that the pulse is not affected by convective mixing (dispersion) after it has passed the photometer. The penetration of a pulse is the volumetric distance between the entrance of the mouth (M) and the mode (concentration maximum) of the bolus at end-inspiration.

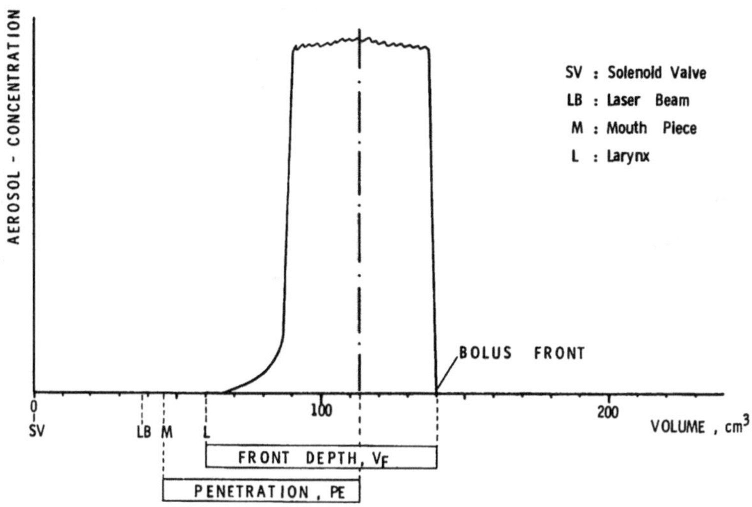

Figure 1: Characteristic parameters of an aerosol bolus

Since the oral cavity represents a large dead space which would cause convective mixing during inhalation it was bridged over by a tube and the cavity outside the tube was filled with a preformed material. So the dead space was reduced to the volume of the tube, pharynx, and larynx. In the example in Fig. 1, the front depth is 80 cm^3, and the penetration 68 cm^3. These values are based on the assumption that the dead space between mouth entrance (M) and larynx (L) was 15 cm^3; this is assumed to be a minimum value, so that for larger dead spaces the front depth becomes even smaller.
The retention measurement of the deposited particles was started immediately after the bolus inhalation.

RESULTS

To bring more light into the results obtained in previous investigtions, experiments were carried out where the size of the inhaled particles and the period of breath-holding were varied in such a way that the settling distance of the particles during the inspiratory pause remained constant. In other words, when the particle diameter, d_{ae}, was increased by a factor of 2, the time of breath-holding was decreased by a factor of 4, so that the sedimentation parameter $d_{ae}^2 \, t_p$ was the same . This procedure was to ensure that at a preset front depth deposition took place in the some volumetric lung depth regardless of the particle size. As a consequence of their equal deposition pattern, these particles should be also cleared in the same manner.

As one can see from Fig. 2, however, clearance was found to depend on particle size in such a way that a greater fraction of the larger particles was cleared in the fast phase, suggesting a smaller deposition in nonciliated airways and a larger deposition in ciliated airways as compared to smaller particles.

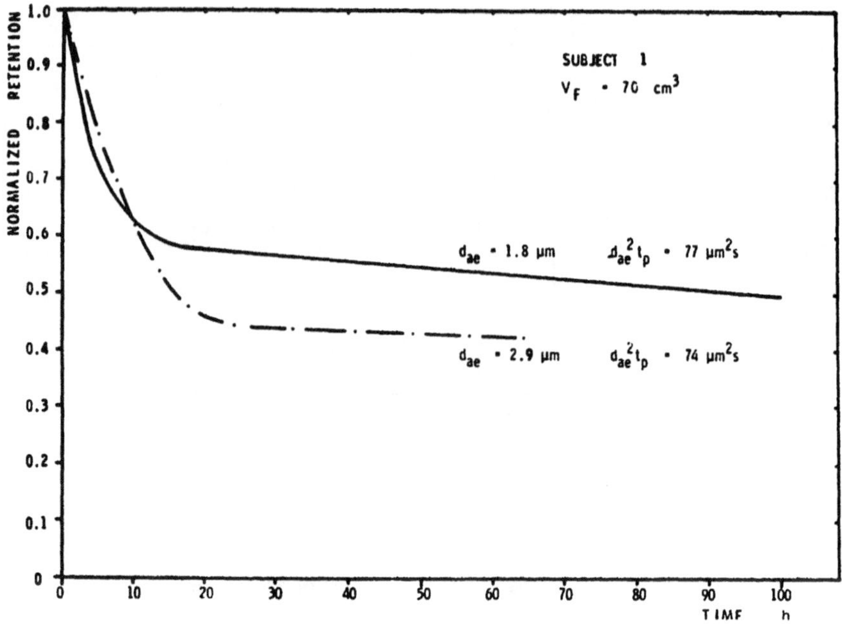

Figure 2: Thoracic retention for the front depth of 70 cm^3 of the inhaled bolus under the condition of equal sedimentation parameter for both curves and different particle size.

SUMMARY AND CONCLUSIONS

There exist several possible explanations for these experimental findings: Firstly, one may suppose that the volumetric depth in which the particles are deposited may be influenced by particle size, either by some size-dependent transport into the lungs, by deposition during inspiration, or by different deposition probabilities during the pause. However, since the sedimentation parameter $d_{ae}^2 t_p$ was kept constant, the sedimentation probability during the pause must be the same at the same lung depth, even for dispersed boluses. Furthermore, since the relaxation time of the largest particles used was only 3 x 10^{-5} sec, inertial forces acting upon the particles during their transport into the lungs can also be excluded.
From bronchial deposition data during steady breathing it can be seen that bronchial deposition of 3 µm-particles amounts only about 3% of inspired aerosol; the corresponding deposition in the pause is estimated to range between about 50% and 70%. Hence, inspiratory deposition may contribute somewhat to the observed effect, but cannot explain the whole phenomenon (2).
Cardiogenic transport influencing particle deposition in the lung leads to another possible explanation. There exists a transport of gases due to cardiogenic motion. West et al. found pulsatile gas flow in the bronchi caused by the heart beat (3,4). Although this flow is bidirectional it leads to an unidirectional transport of particles to the deeper lungs because of the different number of deposited particles for each of the directions. If there is such an effect the 1.8 µm sized particles will reach alveolar structures with a higher

probability than the 3.0 µm sized particles because of their greater breath holding time (see Fig. 2). So the difference of the two clearance curves with the same sedimentation parameter, $d_{ae}^2 t_p$ will be influenced by this phenomenon.

REFERENCES

(1) Stahlhofen, W., J. Gebhart, G. Rudolf and G. Scheuch (1985) Measurement of clearance with pulses of radioactively labelled aerosols, J. Aerosol Sci., 17, 3, 333-336.
(2) Stahlhofen, W., J. Gebhart and J. Heyder (1980) Experimental determination of regional deposition of aerosol particles in the human respiratory tract, Am. Ind. Hyg. Assoc. J., 41, 385-398.
(3) West, J.B. and P. Hugh-Jonea (1961) Pulsatile gas flow in bronchi caused by the heart beat, J. Appl. Physiol., 16, 697-702.
(4) Engel, L.A. (1986) Dynamic distribution of gas flow, Handbook of Ohysiology, Sec.3 (Ed.: P. Fishman), Bethesda, pp. 575-593.

FACTORS INFLUENCING CADMIUM AND LEAD CONCENTRATIONS IN HAIR OF CHILDREN

Wilhelm, M., I. Lombeck*, D. Hafner** and F.K. Ohnesorge
Institute of Toxicology, *Children's Hospital, *Institute of Pharmacology, University of Düsseldorf, Moorenstr. 5, D-4000 Düsseldorf 1, FRG

INTRODUCTION

The use of hair analysis as an indicator of cadmium (Cd) and lead (Pb) exposure of children has not been definitively established (1). Two main problems impair interpretation of trace element concentrations in hair: 1. A distinction between endogenous and exogenous metals cannot be always made. 2. Factors such as sex, age, hair color etc. may cause great interindividual differences in hair metal levels.
To clarify these questions we investigated the factors influencing Pb and Cd concentrations in hair of 474 preschool children living in an industrial town (Duisburg, FRG), and in a rural district (Westfalen, FRG).

MATERIALS AND METHODS

The children underwent a rough medical check-up and body weight, height, hair color and -type were recorded. Data on age, socio-economic status. place of residence as well as on eating habits, smoking habits of household members, nationality and health status were obtained by a questionnaire which was filled out by the parents.
2 strings of hair were clipped as close as possible to the scalp from symmetrical occipital regions. The first 2 cm of the proximal end were used for analysis. The samples were washed in acetone/methanol and digested in quartz flasks using a mixture of nitric-, perchloric- and sulfuric acid. Pb and Cd were determined by atomic absorption spectrometry.
Details of the methods will be published elsewhere (3).

RESULTS

The geometric mean (GM) for Pb in hair of all samples (n=474) amounts to 2.7 µg/g (Range: 0.2-37.7 µg/g) and for Cd to 89.5 ng/g (Range: 10-4281 ng/g). These values are lower than other reported data (3-5). One explanation for this discrepancy could be the different statistical treatment of the results. Otherwise it is known that Cd and Pb levels in hair increase with hair length due to external contamination (6). From many studies it is not clear which part of the hair was used by the investigators.
Table 1 shows the results of the stepwise regression analysis of 23 independent variables on the Pb and Cd content in hair. R means the multiple correlation coeffi-

cient with respect to the variables entered, the increase of RSQ indicates the improvement of the total correlation coefficient by each variable and the F-to-enter values for each variable test the significant improvement of the multiple R.

Table 1 Influence of variables on log-Pb and log-Cd concentrations in hair of 474 preschool children using stepwise multiple regression analysis

Variable	F	R	Increase of RSQ
	log - Pb		
Season	135.3	0.472	0.223
Place of Residence	82.3	0.582	0.116
Sex	76.2	0.656	0.092
Age	25.2	0.678	0.030
	log - Cd		
Season	45.4	0.317	0.100
Sex	26.8	0.395	0.056
Hair color	12.5	0.455	0.025
Place of residence	12.1	0.453	0.024

These factors influence Cd and Pb in hair as follows:
In both study areas the Pb and Cd content of hair samples obtained during summer are nearly 2 to 3 times higher than in winter (Fig. 1). These findings might contribute to the great variation of reported values since most refernces do not take this important variable into account.
With the exception of Cd levels in hair obtained in winter we always found higher metal concentrations in samples of children living in the polluted area compared to those from the rural district (Fig. 1).
Fig. 2 shows that the Cd and Pb levels in hair decrease with increasing age with the most pronounced drop between 3 and 4 years.
In all age groups boys have more Pb and Cd in hair than girls (Fig. 2).

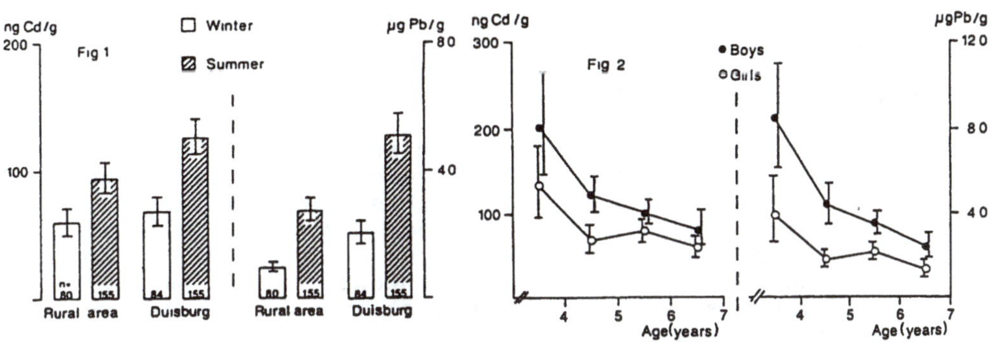

Fig. 1,2: Cd and Pb conc. in hair of children (GM; 95% confid. interv. of mean) in relation to season, place of residence (Fig. 1) and to sex, age (Fig. 2).

Furthermore, in Duisburg hair-Cd and Pb of children whose fathers are factory workers are higher compared to other occupations (values are given as gm; Cd:112.3 vs. 90.6 ng/g; $P<0.05$; Pb: µg/g; Pb:4.2 vs. 3.0 µg/g; $P<0.01$).
We suppose, that the influence of the variables season, sex, place of residence, and age are associated with the playing habits of the children. By playing outdoors children take up Pb and Cd contaminated particulates deposited in air, dust, and soil. The ingestion rate by hand to mouth activities decreases with increasing age thus explaining the relationship between metals in hair and age dependency. As shown by Ewers and coworkers (7) home contamination with industrial dust can contribute to an increased exposure of household members which could account for the influence of father's occupation. Looking at the data on environmental pollution (8), we found a positive relationship between Cd in dustfall and Cd in hair obtained from boys during summer aged 3 to 4 years ($R=0.44$). Pb content in soil is related to hair Pb levels of all children from Duisburg ($R=0.34$; $P<0.001$). As the oral absorption rate for Pb and Cd for children is thought to be much higher compared to adults the ingestion of contaminated soil in this age group may greatly contribute to the whole Pb and Cd uptake.
Apart from environmental conditions we found that hair Cd concentrations decrease with the hair color from red to blond to black. These observations might be related to the melanin content of hair which follows an inverse pattern when red hair is excluded.
Considering hair Pb and Cd levels of all children, no impaired health status according to the exposure was observed. But the log Pb and log Cd concentrations are inversely related to the zinc (Zn) levels in hair (log Pb µg/g vs. Zn µg/g: $y=-0.0056x+1.1$; $R=-0.593$; $P<0.001$; log Cd ng/g vs. Zn µg/g: $y=-0.0034x+2.36$; $R=-0.436$; $P<0.001$).
Factors such as nationality, smoking habits of household members, eating habits, duration of residence, general health status, and anthropometric data did not show any relation to Cd and Pb concentrations in hair of the children.

SUMMARY AND CONCLUSIONS

The main influencing factors on Pb and Cd in hair of children are season, sex, and place of residence. No correlation between an impaired health status and the Pb and Cd content in hair was detected. There exists an inverse relationship between Pb and Cd on one hand and Zn on the other hand. Since the Zn content in hair is well recognized to reflect nutritional status and the external Zn contamination of the hair is negligible we conclude that our findings concerning Pb and Cd in hair are not impaired by external contamination and furthermore that for estimating the health risks of toxic metals by hair analysis the Zn status should be included.

ACKNOWLEDGEMENTS

The authors thank Mrs. B. Obertrifter, Mrs. P. Breuer, Mrs. K. Stöcker and Mr. H. Ohnesorge for their skillful technical assistance. The support of the Bundesministerium für Jugend, Familie, Frauen und Gesundheit, Bonn, FRG, is greatfully acknowledged.

REFERENCES

(1) Krause, C. and M. Chutsch (eds) (1987) Haaranalyse in Medizin und Umwelt, Schriftenreihe des Vereins für Wassser-, Boden- und Lufthygiene 71, Gustav Fischer Verlag, Stuttgart/New York

(2) Wilhelm M., D. Hafner, I. Lombeck, and F.K. Ohnesorge, Variables influencing cadmium concentrations in hair of preschool children living in different areas of the Federal Republic of Germany, Int. Arch. Occup. Environ. Health, in press.

(3) Barlow P.J., P.E. Francois, I.J.L. Goldberg, I. Richardson, M.G.A. Izmeth, K. Kumpeson and P. Sykes (1986) Trace element abnormalities in long-stay hyperactive mentally handicapped children and agitated senile dements, J. R. Soc. Med., 79, 581-583.

(4) Wibowo, A.A.E., R.F.M. Herber, H.A. Das, N. Roeleveld, and R.L. Zielhuis (1986), Levels of metals in hair of young children as an indicator of environmental pollution, Environ. Res., 40, 346-356.

(5) Wecker, L., S.B. Miller, S.R. Cochran, D.L. Dugger, W.D. Johnson (1985) Trace element concentrations in hair from autistic children, J. Ment. Defic. Res., 29, 15-22.

(6) Bos, A.J.J., C.C.A.H. van der Stap, V. Valkovic, R. D. Vis, and H. Verheul (1985) Incorporation routes of elements into human hair: implications for hair analysis, Sci. Total. Environ., 42, 157-169

(7) Ewers, U., R. Dolgner, A. Brockhaus, I. Freier, E. Jermann, W. Hilser, K. Domisch, and R. Braun (1984) Untersuchungen zur Blei- und Cadmiumbelastung von Kindern und Jugendlichen in einem schwermetallbelasteten Gebiet (Stolberg), Öff. Gesundh.-Wes., 46, 231-236.

(8) Landesanstalt für Immissionsschutz des Landes Nordrhein-Westfalen (1985) Schriftenreihe 63, Verlag W. Girardet, Essen.

IV. EPIDEMIOLOGY

CYTOGENETIC FINDINGS IN STYRENE WORKERS IN RELATION TO EXPOSURE

Forni, A., E. Goggi°, E. Ortisi, R. Cecchetti°, G. Cortona°°, G. Sesana°° and L. Alessio°°°
Institute of Occupational Health, University of Milan, V. S.Barnaba 8, 20122 Milan, Italy; °Unit of Occupational Health, USSL 61, Carate Brianza (Milan); °°Service of Occupational Medicine, USSL 63, Desio Hospital, Desio (Milan); °°°Institute of Occupational Health, University of Brescia, Brescia, Italy

INTRODUCTION

Styrene is a volatile fluid which is increasingly used in the production of polyester resins due to its capacity to polymerize. Since styrene and especially styrene oxide, one of the styrene metabolites in humans, have resulted to be mutagenic and clastogenic in several short term tests, a possible carcinogenic effect of this solvent has been suggested (1,2). In the last decade, several cytogenetic studies have been performed in groups of workers exposed to styrene in different industrial settings with conflicting results (see reviews in 1,3).

The aim of the present investigation was to evaluate the possible cytogenetic effects of styrene in two groups of workers with different degree of exposure, comparing the results of chromosome aberration studies with those of matched controls, and relating the results to the exposure data, which were available for all workers for many years.

MATERIALS AND METHODS

Subjects and Factories

The styrene workers investigated were employed in two factories of the Milan hinterland, with different characteristics. In factory A, a medium size manufacture of reinforced plastic laminates and insulating polymers, which is operating since a long time, the atmospheric concentrations of styrene had been rather high up to 1978, when they have been drastically reduced by automation of the production cycle (Table 1). In factory B, a rather recent small plastic boats manufacture, the styrene concentrations vary during the year according to the size of the boats produced (Table 1). In both factories, the only toxic used, beside styrene, was acetone, which was used in small quantities for cleaning the machineries.

The subjects studied were 40 male styrene workers and 40 external controls matched for age ($\pm$ 5 years) and smoking habits, living in the same industrialized area. All subjects were healthy and had no history of exposure to other known chromosome-damaging agents (radiation treatment, occupational exposure to other known clastogens, prolonged intake of drugs, recent viral diseases). Workers of factory A were older than those of factory B (mean age 45.2 $\pm$ 5.8 versus 27.9 $\pm$ 5.6) and had been exposed for longer times (see Table 1). The current exposure was much lower for workers of factory A than for those of factory B, as shown in Table 1 and documented also by the urinary excretion of mandelic acid (data not shown). For each exposed subject, a cumulative exposure index was calculated by multiplying the atmospheric concentrations of styrene at the different worksites by years of exposure, taking into account also changes of job; this index, however, may represent an underestimation, since also absorption through the skin might have occurred especially in the past, in some jobs. For workers of factory A, most exposure had been cumulated before 1978.

Table 1. Exposure data of styrene workers

Groups	Years of exposure $\bar{x} \pm$ SD	TWA styrene mg/cum° up to 1978	after 1978	Cumulative exposure°° $\bar{x} \pm$ SD
Factory A 32 workers	18.8 ± 3.9	123 - 249	1.7 - 17	1077 ± 715
Factory B 8 workers	4.5 ± 3.4	——	41 - 198	471 ± 344

° range of mean values at different worksites
°°air styrene mg/cum x years of exposure

Cytogenetic Methods

The chromosome aberration studies were carried out from June 1985 to May 1986. Peripheral blood lymphocytes were cultured for 2 days according to the method in use in the laboratory (whole blood in TC Medium 199 Difco + 10% autologous plasma + phytohemagglutinin M Difco). For each subject 100 well spread metaphases were counted and scored for structural chromosome aberrations, according to Buckton and Evans and to ISCN (4,5). The rates of total abnormal metaphases (including and excluding gaps), of cells with chromatid-type aberrations (chromatid breaks and exchanges, B cells) and of cells with unstable chromosome-type aberrations (acentric fragments, dicentric and ring chromosomes, C_u cells) were calculated separately for the workers of the two factories, and statistically compared with those of the matched controls.

RESULTS

The results of the chromosome studies are summarized in Table 2.

Table 2. Rates of chromosome aberrations in styrene workers and in the matched controls (mean values)

	Subjects	Total cells studied	% Abnormal metaphases including gaps	% Abnormal metaphases excluding gaps	% Cells B	% Cells C_u
Factory A	Workers	3200	3.2	2.3	1.1	1.2
	Controls	3200	2.9	1.6*	1.25	0.3**
Factory B	Workers	800	4.25	2.5	2.25	0.25
	Controls	800	2.9	1.5	1.0*	0.5

* $p < 0.05$ ** $p < 0.001$ at the Wilcoxon matched-pairs test
B: cells with chromatid breaks or exchanges
C_u: cells with acentric fragments, dicentric or ring chromosomes

In workers of factory A, the rate of total abnormal metaphases excluding gaps was significantly higher than in the matched controls at the Wilcoxon matched-pairs test (T = 91 for 26 pairs, p = 0.016), and the difference was highly significant for chromosome-type unstable aberrations (T = 7 for 20 pairs, $p < 0.001$). Chromosome-type stable aberrations were very scanty both in the exposed and in the controls. The distribution of the exposed and controls for number of chromosome-type unstable aberrations and of asymmetric chromosome exchanges (dicentric + ring chromosomes) differed significantly at Fisher's exact probability test (see Table 3).

Table 3. Distribution of styrene workers of factory A and of the matched controls for rates of unstable chromosome-type aberrations (C_u) and of asymmetrical chromosome exchanges

	% C_u cells		% dicentrics + rings	
	< 2	≥ 2	0	≥ 1
Workers No.	20	12	24	8
Controls No.	32	0	30	2
Fisher's test	$p < 0.001$		p = 0.003	

Even though no correlation was found between exposure indices and chromosome aberration rates, the group of subjects with higher rates of chromosome-type aberrations or with dicentric and ring chromosomes had higher mean values of cumulative exposure. In fact, the mean cumulative exposure of the 12 subjects with C_u cells $\geq$ 2 % was 1067, that of the 15 subjects with C_u cells $\geq$ 2 % or at least 1 % dicentric or ring chromosomes was 1217, that of the 8 workers with at least 1 % dicentric or ring chromosomes was 1763 and no lower than approximately 1000.

In the 8 workers of factory B (Table 1), the cytogenetic study showed an increased frequency, compared to the matched controls, of chromatid-type aberrations (breaks + exchanges) at the Wilcoxon matched-pairs test (T = 2.5 for 7 pairs, p = 0.026). Gaps were not significantly increased. No chromosome-type stable aberrations were observed.

SUMMARY AND CONCLUSIONS

The results of the present study confirm an association between styrene exposure in the production of reinforced polyester resins and increased chromosome aberrations in cultured lymphocytes, in agreement with most reports of the literature (see reviews in 1-3). The possible influence of age and smoking on chromosome aberration rates in this study are ruled out by the strict matching of controls for these variables.

The fact that increased rates of chromosome-type aberrations and of dicentric or ring chromosomes, compared to the matched controls, were found only in workers of factory A, who had very low current exposure at the time of chromosome studies, but had had high and long-term exposure in the past with consequent high cumulative exposure, suggests a persistent cytogenetic effect of the past high-level exposure.

Workers of factory B, who had higher current exposure than those of factory A, but low cumulative exposure, presented only an increase of chromatid-type aberrations. The results suggest the opportunity of a follow-up of these workers, in order to study the possible evolution with time of chromatid-type aberrations into "derived" chromosome-type aberrations in vivo. Should this hypothesis be confirmed, industrial hygiene preventive measures should be suggested, even though the exposure is still within limits currently considered safe.

Supported in part by a grant of the Ministry of Education (MPI 60%) to A. Forni.

REFERENCES

(1) International Agency for Research on Cancer (1982) Chemicals, Industrial Processes and Industries associated with Cancer in Humans, IARC Monographs on the Evaluation of the Carcinogenic Risk of Chemicals to Humans, Suppl. 4, IARC, Lyon, 229-233.
(2) Lauwerys, R. (1984) Styrene, in L. Alessio, A. Berlin, M. Boni and R. Roi eds., Biological Indicators for the Assessment of Human Exposure to Industrial Chemicals, EUR 8903 EN, Commission of the European Communities, Luxembourg, 67-81.
(3) Ashby, J. and C.R. Richardson (1985) Tabulation and assessment of 113 human surveillance cytogenetic studies conducted between 1965 and 1984, Mutat. Res., 154, 111-133.
(4) Buckton, K.E. and H.J. Evans eds. (1973) Methods for the analysis of human chromosome aberrations, WHO, Geneva.
(5) ISCN (1978) An International System for Human Cytogenetic Nomenclature, Cytogenet. Cell Genet., 21, 309-404.

INFLUENCE OF AIR POLLUTION TO RESPIRATORY DISEASES IN INFANCY

Haupt, H.
Städt. Kinderklinik Duisburg, Zu den Rehwiesen 9,
4100 Duisburg 1, FRG

INTRODUCTION

Our examination concerns with a possible relation between a high air pollution and the frequency of illness with subglottic laryngitis and spastic bronchitis in childhood. In case of subglottic laryngitis is observed an inspiratoric stridor and inspiratoric dyspnea. Both diseases are typical for babies and young children. They are caused by viruses or more rarely by bacteries. Air pollution could only be able to aggravate the situation.

MATERIALS AND METHODS

In a retrospective study we have investigated children who had been stationary treated in our children's hospital in the years 1979 till 1982. For this we observed the frequency of illness on the one hand on places of dwelling with different immissions, on the other hand during days of different air pollution. Measured and valued have been the concentrations of sulphur dioxide, nitrogen oxide, nitrogen dioxide and dust.

The first examination took place in an area with lateral length from 6 and 10 kilometres, alltogether 60 square kilometres. The position of our hospital was located in the centre of this area. Therefore the longest distance to our hospital was smaller than 6 kilometres. The stronger loading regions were more remoted to the hospital than the smaller loading regions. Therefore it was not possible that an incorrect positive correlation between immission and frequency of illness was simulated through the different distances to the hospital.

It is possible that the social state has an influence on the result of the examination. Therefore the density of population, the part of foreigners and the mode of heating have been considered. This examination shows that the social state is the best represented through the children density and the part of foreigners.

RESULTS

The frequency of illness was greater in areas with a high immission than in areas with a small immission. The social state has not been important to these findings. In areas with a high or small part of foreigners was equally found, that on high immission the frequency of illness was larger than in areas with a small immission. The frequency of illness equally has been dependent on the amount of the immission in foreign and german children.*) Also the density of population did not show a significant influence on the frequency of illness (1, 2, 3). *) (Tab.1)

A control by a prospective study in the years 1983 and 1984 at the same mode of investigation shows the same findings although the immission than has been decreased.

The examination on a temporal correlation between air pollution and frequency of illness had been executed only in the winter halfyear from October to march. There the air pollution of single days had been compared with the frequency of illness on these same days.

In these examinations meterological parameters as temperature, atmospheric pressure and humidity, also weather conditions (according to Ungeheuer) have been considered (Tab.2). There was not found a significant correlation between meteorologic parameters and frequency of illness.

But also in this study the frequency of illness increased, simultaneously with increasing air pollution (4, 5).

On days following days with a high air pollution the frequency of illness was still increased.

The frequency of illness increased more, if more than one substance was elevated (Fig. 1) (1,2,4,5).

To that study also a prospective control study was executed. From the town of Duisburg in a district with a high immission a measuring station was installed. From the pediatrician established in this area we got the number of cases of illness and their beginning on every day. Also in this control study we found a correlation between the degree of air pollution with sulphur dioxyde and the frequency of illness. Dr. Schneider and Dr. Struckmeier of the chemical institute of Duisburg have calculated and found a significant correlation between the values in the air from sulphur dioxide and dust. This finding is also able to attest that the air pollution with some agents simultaneous causes an increase of the frequency of illness.

All our referred studies show a correlation between the immission in places of dwelling and on the seperate days on the one hand and the frequency of illness on the other hand, corresponding with a chronic and an acute load by air pollution.

Threshold values have not resulted by our methods. Also it was not possible to hold one single agent in the air responsible for the refered effect. But each of the measured substances represent an indicator for the air pollution and their influence on the health.

SUMMARY

Our examination concerns with the question of a possible coherence between a high air pollution and the frequency of illness with spastic bronchitis and subglottic laryngitis in childhood. For this we observed the frequency of illness, on the one hand in places of dwelling with different immissions and on the other hand at times with different air pollution. There were available measured values of SO_2, NO, NO_2 and dust. By doing so the known interference factors were considered. Particularly the social status of the concerning families and meteorologic influences were examinated. It was found out that the frequency of illness increased in regions and at times with a great air pollution. The frequency of illness increased stronger if the concentration of more than one substance was elevated. Therefor it is not to demonstrate that any especially agent is responsible by itself for the increase of the frequency of illness.

REFERENCES

(1) Mühling, P., J. Bory, H. Haupt: Einfluß der Luftbelastung auf Atemwegserkrankungen des Kleinkindes. Fortschr. Med. 102, 831 - 834 (1984)

(2) Mühling, P., J. Bory, H. Haupt: Einfluß der Luftbelastung auf Atemwegserkrankungen. Staub - Reinhaltung Luft 45, 35 - 38 (1985)

(3) Haupt, H.: Pseudokrupp, Duisburger Studie in E.v. Nieding u. K. Jander, Umwelthygiene. Schriftenreihe des Vereins. Wasser-Boden-Luft-Hygiene Nr. 65, G. Fischer, Stgt.-N.Y. 1985. S. 77 - 79

(4) Haupt, H.: Krupp-Studie Duisburg in E.v.Nieding u. K. Jander. Smogepisoden. Schriftenreihe d. Vereins. Wasser-Boden-Luft-Hygiene Nr. 69, G. Fischer, Stgt.-N.Y. 1986, S. 159 - 195

(5) Haupt, H. u. J. Bory: Duisburger Pseudokrupp - Studie in: Aktuelle Probleme der Luftreinhaltung: Pseudokrupp, Dioxine/Furane. Schriftenreihe d. VDI-Kommission. Reinhaltung d. Luft Bd. 2, 7 - 12 (1986)

	SO_2 (I_2) mg/m^3	cases of laryngitis per 100 children	factor	p
german children	≤ 0,3 > 0,3	10,3 20,2	+ 1,96	< 0,001
foreign children	≤ 0,3 > 0,3	9,7 19.5	+ 2,00	< 0,01

	dust (I_2) $g/m^2.d$	cases of bronchitis per 100 children	factor	p
german children	≤ 0.55 > 0.55	10,6 26,9	+ 2,54	< 0,001
foreign children	≤ 0.55 > 0.55	22,7 37,6	+ 1,66	< 0,001

Tab. 1

weather conditions	biological	cases of laryngitis per 100 days
1, 2 } fair weather 6 weather quiting	suitable	36,8
4, 5 } shift of the weather 6_z invasion of cold air	unfavorable	36,5

Tab. 2

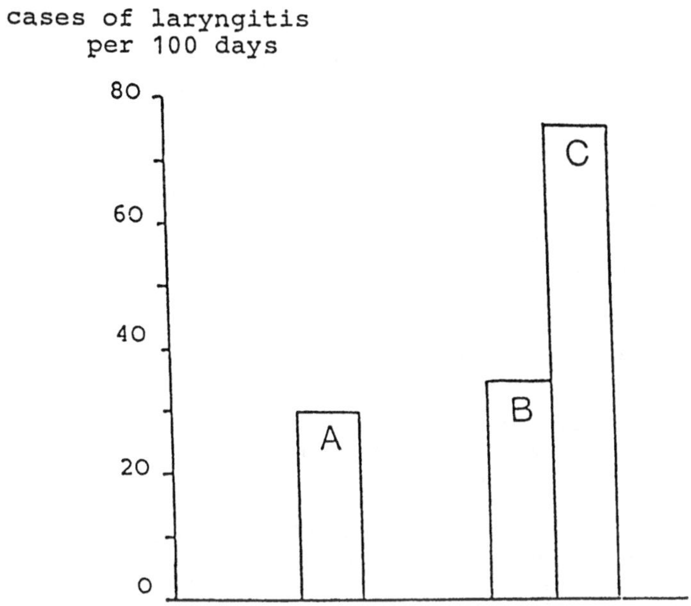

A: all agents below limit
B: one agent above limit
C: some agents above limit

arbitrary limits (I_2):

SO_2, NO and dust = 0,3 mg/m^3
NO_2 = 0,1 mg/m^3

Fig. 1

TOXICOPY - A BASIC MECHANISM TO COPE WITH ENVIRONMENTAL THREATS

Kofler W. W. (Department of Social Medicine at the University of Innsbruck, Sonnenburgstrasse 16, A-6020 Innsbruck, Austria

INTRODUCTION

In the sphere of immission of a quartz-sand-plant symptoms of illnesses not unlike the ones normally occurring with poisoning could be registered (7, 10). All persons with symptoms of poisoning who were examined, indicated that they felt threatened by poisons of unknown kind caused by the quartz-sand-plant. Relevant poisons in even roughly relevant concentrations were not found.

Examples for Observed Phenomena and other Facts which cannot sufficiently be explained

Mass diseases being attributed to poisons by the affected ones only are described in literature several times. Especially COLLIGAN and his team at NIOSH report mass hystery, psychogenic mass diseases, production-line-hystery etc. In spite of great investigation-efforts no relevant environmental stress could be proved. COLLIGAN and his co-workers can work out remarkable epidemiologically common properties, as there are high psychosocial stress under working conditions with perceivable environmental factors, especially smell substances, working as triggers in consequence (1, 3, 4).

The phenomenon which is best investigated, is the placebo-phenomenon. In the double-blind-study there can be obtained effects of poisons without any pharmacological substance as long as the patient believes in a treatise with a pharmakon.

In connection with environmental disasters, e. g. Chernobyl, symptoms have partially been registered with many persons who were not exposed to the relevant environmental toxins (6). Common to all affected ones was the feeling of a threat by an unphysiological substance. According to the present level of knowledge threats are typical stressors, thus causing reactions in the sense of the General Adaptation Syndrome. Although such reactions could have been expected, they are not reported in any of the cited works (4).

Some Principles of the Additional Working Theory to the Theory of Coping Strategy

For the enlargened model of strategy of coping with environmental threats the following premises in essential correspond to basical biological experience:

1. Where the amount of information is concerned, perceptible factors of the environment are evaluated by the achievements of brain. The resulting concious of unconscios assessment of the situation can be manifold. Threat is only one of the many possibilities.
2. The brain can steer the peripheral organs consciously or unconsciously. In this process all existing steering mechanisms are integrated (medicated by the hormones or the nerves?). The manner of this integration, especially the step from making a decision to carrying it out and the formation of somatical reactions has not yet been sufficiently studied. This is true for conscious steps as well as for unvoluntary ones.
3. Man has proven to be a living being especially suitable for adap-

tation. Any assumption of his unability to cope with important dangers of long durance and high frequency would contradict the ruling teaching positions and thus require well founded argumentation.

4. The mobilisation of all forces to withstand a danger by flight or application of physical force is an adequate mechanism in case of a threat of the physical existence by being slain, eaten or crushed to death, or something of the kind (existential threat).
5. Food which has been rendered poisonous e. g. by processes of decaying, primary poisonous of infected food, choking smoke or similar unphysiological substances have been everyday threats for man since thousands of years. They cannot be coped with by attack or mobilisation of energy (substantial threat). Much rather, the ability to differentiate between food that is just about to be edible and food that cannot be eaten any more as well as the ability to sufficiently react in cases of an uptake of poison in spite of differentiation, must have been even more valueable than today. In former times, not to forget, famines were not rare and the possibilities of obtaining food were very limited, so that those abilities may have been deciding advantages for surviving. As man is not able to directly recognize poison, poison threats have to be deduced from informations the existing sense-organs in combination with intellectual operations provide. Especially the chemical senses and deductions from observations of the behaviour of members of the own species, the own body and from believable indications, respectively, are of overwhelming importance.
6. As a further kind of threatening the threatening of man's psychosocial integrity has to be regarded. This one as well cannot be adequatly coped with but only by social-communicative coping mechanisms.

Toxicopy as Part of the Mechanism of Coping with Threats (Working Hypothesis)

The expression "Toxicopy" has been chosen by KOFLER for the occurrence of the symptoms of poisoning in absence of the relevant poisons: the state of poisoning is copied to cope with a suspected threat of poison as adequately as possible.

The following hypothesis have been developed with regard to principles and different models (2, 11) which in any case can only partly be adopted. We did not test the single steps experimentally, as an isolated group of scientists could hardly do. Still we tested accessible cases of psychogenic mass diseases what their contradiction to our explanation is concerned. Up to now we have not been able to find any serious contradiction. Additionally, based on those hypothesis changements of behaviour, persistance and disappearance of symptoms, which neighbours of the quartz-sand-work showed, could be predicted when a cleaning device was prescribed. Though we have not been able to termine investigations yet, we can already make several statements: persons who where told foreward to be annoyed even after reduction of eventually noxious emission, showed symptoms when emission was far below the level it was when nobody felt sick yet.

It seems to be justified to put our hypothesis to public discussion even if it is relatively weakly proven yet: the matter may be highly important for our understanding of the environment. It is practically applicable in a sphere which is lacking generally applicable starts.

Preliminary Phase

1st step: An unknown perceptible factor of the enviroment or one that is supposed to be dangerous leads to a raise of attention thus lowering of the level of awaking. At the same time the readiness to withdraw by flight, if necessary, is raised.
2nd step: In this step a first evaluation of the content of information makes it clear whether a coping strategy in the very place is advised or an immediate flight (under spending of all available forces) seems preferable.
3rd step: This next step is a differentiated evaluation of the information context to know whether the situation really represents a threat or, in fact, a completely different evaluation is reasonable (e. g. positive situations like joy, need of help, irrelevancy of the information etc.).

Primary Threat Management (Table 1)

4th step: In case a threat cannot be excluded, in a further evaluation the nature of the threat and the probability of an assumption of one or more kinds of threats, respectively, are assumed. Therefore (at least) three possibilities can be chosen from: existential threat, substantial threat or threat of psycho-social integrity.
5th step: After this assignment we principally can choose between two alternatives. It is possible to either apply one of the relevant adequate coping mechanisms or to withdraw from the threat by flight (dislocalisation).

Alternative: Coping by Staying-in-the-same-Place Applying one of the Relevant Coping Mechanisms

If leaving is impossible or improper there will have to be chosen coping mechanisms allowing to stay in the same place: adequate in case of existential threat will be the mechanism of stress, in case of threat of the psycho-social integrity it will have to be socially communicative coping strategies, in case of threatening by toxins - toxicopy.

The principle of toxicopy is the fact that man does not know any mechanism of detoxication that can protect him better than not taking the toxin up all, or at the utmost in tiny amounts. This is also true for substances that are only supposed to be toxic, which is essential for our theory. To initiate the most effective mechanism of protection, the not-taking-up, only intellectual efforts can be decisive. The initiative does not at all depend on the real existence or non-existence of the poison. In case the uptake of poison cannot be excluded it will be advised to increase excretion to a maximum as long as there are no specific patterns of reaction, or they are unknown, respectively. At the same time metabolism is reduced as much as possible to be no longer dependent on the uptake of products that are suspected to be poisonous.

In addition to disturbances of sleep (as a consequence of a lowering of the level of awaking) table 2 shows the symptoms of unspecific toxicopy. This unspecific pattern of reactions proves to be uneconomic. It is replaced by specific reactions as soon as the affected one believes to be able to deduce adequate reactions from observations. The threat of the reactions results in specific symptoms exclusively. Transferred to environmental hygiene, this signifies the possibility of creation of practically all pictures of intoxication as long as the presence of the accused substance is not required (e. g. brown maculae of fluorosis only in presence of fluor). Thus teratogenic effects are as little expected as mutagenic ones. The well known phenomenon of placebo-effect can be enumerated as classical example of a specific toxicopy: the medical doctor has to provide the information of the effects of the "medicament" on the body to "make" the body of the patient react.

The presented model is most possibly able to explain the symptoms in the neighbourhood of our quartz-sand-plant, the illnesses in Cis-Jordany and other school-epidemies as well as give a theory free from contradictories explaining the placebo-phenomenon which could not be made transparent yet. Still there are no explanations for most of the psychogenic mass diseases in working-places described by COLLIGAN et al. Also it is not sufficiently evident why mechanisms of stress are chosen without existential threat.

This wants an enlargement of the model:

Secondary Threat Management (Table 3)

The key-functions for all situations which are regarded to be dangerous lies in the evaluation of the processes of thinking. Thus in this sphere too have to be assumed processes of weighing and examining. Primary aims are headed for primarily thus not allowing mechanisms which are likely to endanger those aims. They would not be allowed even if they were estimated adequate for coping with the secondary problem. Adequate coping mechanisms could also not be applied if, in the situation, they were not applicable (e. g. absence of a possibility to escape, lacking partner of communication because the person in question has not learned to apply the method suitably, e. g. solvation of problems in partnership by communication). In all those cases the appliance of the adequate coping strategy would not be compatible with the real coping-capacity.

As well known from the theory of neuroses similar situations are not answered by non-action but by changement to a different, permitted strategy of coping. Most probably, processes of attribution occupy a key-function to help the individual to believably change his interpretation of the situation in a way that can initiate deviating-reactions.

In this way reactions of unskilled female workers in shoe- and fish-packing-factories (4) could be explained. Being underpayed, standing under extreme psycho-social stress which could not be coped with by social-communicative methods because of the threat of dismission, they had once noticed the ever present smell of glue or fish, respectively. As they regarded it to be a threat they responded by falling ill. Headache, nausea, difficulties with the circulatory system etc. became symptoms spreading within the plants like an epidemy. Regarding data of the last years, a change to mechanisms of toxicopy in absence of a perceivable trigger appears however to be improbable. In our achievement-oriented society it is far more probable that people switch to the generally accepted and proven patterns of coping with stress. Thus secondary stress-reactions arise as well as secondary toxicopy-mechanisms. It is very improbable that the real causes, the threat of psycho-social integrity is thus coped with. In the course of time it will necessarily come to an exhaustion of the coping capacities present and consequently to symptoms like a rise of the blood-pressure and of the rate of coronary infarct.

Regarding the very limited time we could present the theories of primary and secondary threat management only in its main traits. At the moment it is being examined in the frame of a transcultural investigation programme what its consistence with existing theories and present result of investigations is concerned. Until now no incompatibility could be stated.

SUMMARY

Especially US-scientists have described epidemies occurring at work places (1, 3, 4) as well as in schools (9), where no relevant poisons could be proven although the illnesses seemed to be poisonings. The authors deduced those illnesses which they assumed to be widespread, from psychogenic mechanisms. There did not exist any medical explanation or working hypothesis that agreed with nowaday's models of etiology and pathogenity.

There are also no explanations for the fact that the described stress never led to reactions that could have been expected according to the General Adaptation Syndrome of Selye, but in most cases to unspecific poisoning symptoms. Lercher and Kofler have now described similar psychogenically caused mass diseases in the community setting (7, 10).

There has been constructed a working hypothesis to explain the non-occurrance of symptoms that had been expected but did not occur. The explanation should agree with existing theories as far as possible. Therefore it was necessary to change the General Adaptation Syndrome to have the phenomenon regarded as an adequate mechanism for certain evaluation situations but not for all of them. While stress is regarded to be an adequate coping mechanism in cases of physical threat, it is not adequate in other cases. Threat by substances, like poisons or rotten food, radio-active radiation, situations when the self-understanding or the social position is regarded to be threatened, are examples for life-situations where stress does not appear to be adequate.

Social-communicative coping-mechanisms are regarded to be adequate in situations where the psycho-social integrity is threatened. In case of threatenings by poisons specific and unspecific toxicopies replace other coping mechanisms. There are cases in which no adequate method of coping can be applied, either because it has not been larned, has remained without success, is contradictory to other, maybe higher aims or something of the kind. This can lead to some kind of shifting to other methods of coping thus creating secondary toxicopies, secondary stress etc. The secondary reactions will hardly ever really solve the underlying problems. Duration of the state provided, they necessarily will lead to illnesses of some kind.

"Toxicopy" signifies the occurrance of symptoms of poisoning in absence of the relevant amounts of poison. We understand those symptoms as prophylactic countermeasures of the body. It takes up as little poison as possible and / or tries to get rid of the one it may have taken up by chance (= unspecific toxicopy). If excretion is impossible or in addition to it there are chosen reactions that are regarded to be able to cope with the poisoning that threatens

(= specific toxicopies). Toxicopies are thus limited to reaction-patterns and ranges for which the presence of the feared poison is not necessary. That explains the absence of stains of fluoride on teeth in case of a fluoride-toxicopy. The symptoms stay upright only as long as the threat is supposed subjectively. Accepting our hypothesis one could not only explain the phenomenon of psychogenic mass diseases but also the one of placebo. The latter we can understand as specific toxicopy. With its help we could correctly foresee which way neighbours would react on the installation of an odour-cleaning equipment what the persisting and disappearing of symptoms is in case of a toxically irrelevant odour-emittant.

We regard the toxicopy-mechanism as a human answer to threat. It is especially an answer to threat by rotten food which becomes the more relevant the greater the lack of food is. It makes the integration of intellectual abilities (processes of learning, experiences etc.) at the evaluation of still edible food possible and makes it easier to cope with poison that has been taken up erroneously.

As chemistry plays such a great role in our environment even harmless concentrations of a substance may initiate this mechanism as a consequence of inadequate subjective evaluation.

References

(1) Boxer PA: Occupational mass psychogenic illness. J. Occup. Med. 27, 12, 1985

(2) Cannon WB: Stress and strains of homeostasis. Am. J. Med. Sci., Vol. 189, 1935, pp. 1 - 4

(3) Colligan MJ, Murphy LR: Mass psychogenic illness in an electronics plant. J. Occup. Med. 26, 5, 381 - 5, 1984

(4) Colligan MJ: Psychogenic mass diseases and other unclear epidemics on work places: Results of NIOSH and other institutes. Proceedings of the 1st Workshop on Toxicopy, Bangkok 1987, in press.

(5) Henry JP, Stephens PM: Stress, health and the social environment: A sociobiologic approach to medicine. Springer-Verlag

(6) Kofler W: Editorial. Staub-Reinhaltung der Luft, 46, 9, 1, 1986

(7) Kofler W: Toxicopies caused by odours: A form of a general mechanism of protection against ecological damage. Proceedings of the 7th World Clean Air Congress, Sydney, August 1986, Vol. V, pp. 215 ff.

(8) Kofler W: Toxicopy and other support mechanisms of threat management. Proceedings of the 1st Workshop on Toxicopy, Bangkok 1987, in press.

(9) Lercher P: Mass psychogenic illness in the non-occupational setting: Epidemiologic considerations. Proceedings of the 1st Workshop on Toxicopy, Bangkok 1987, in press.

(10) Lercher P, Kofler W: Odour caused mass illness around a plant producing quartz and feldspar by flotation. Proceedings of the First European Meeting of Environmental Hygiene, Düsseldorf, May 21 - 22, 1987, in press.

(11) Selye H: Present status of the stress concept. Clinical Therapeutics, 1, 1, 3 - 19, 1977

TABLE 1: PRIMARY THREAT MANAGEMENT

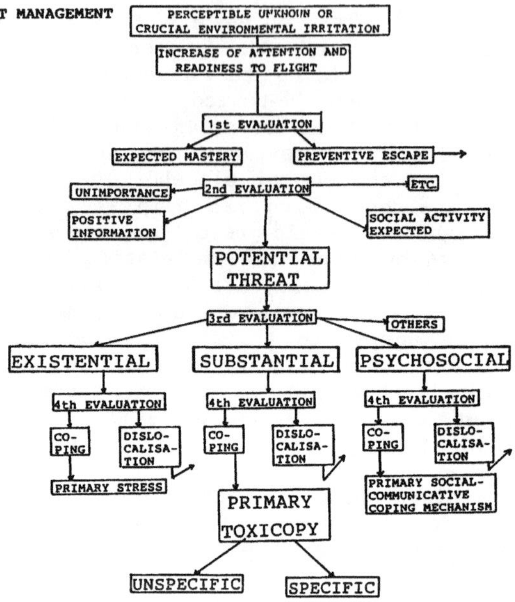

TABLE 2: PREVALENCE OF SYMPTOMS REPORTED IN COMPARISON WITH OTHER EPISODES OF MASS PSYCHOGENIC ILLNESS

THIS STUDY		SUMMARY OF 16 NIOSH INVESTIGATIONS *)	NO. REPORTED
HEADACHE	65,7 % (44)	HEADACHE	14
THROAT IRRITATION	65,7 % (44)	DIZZINESS/LIGHT-HEADEDNESS	12
EYE IRRITATION	53,7 (36)	NAUSEA	11
NOSE IRRITATION	52,2 % (35)	DRY MOUTH/THROAT	10
RESPIRATORY IRRITATION	52,2 % (35)	EYE/NOSE/THROAT IRRITATION	9
NAUSEA	37,3 % (25)	SLEEPINESS/DROWSINESS	7
UNSPECIFIC FEELINGS OF THREATENING	37,3 % (25)	WEAKNESS	7
		NUMBNESS/TINGLING	6
MOUTH IRRITATION	19,4 % (13)	CHEST TIGHTNESS/DISCOMFORT	6
SKIN IRRITATION	10,4 % (7)		

TABLE 3: SECONDARY THREAT MANAGEMENT

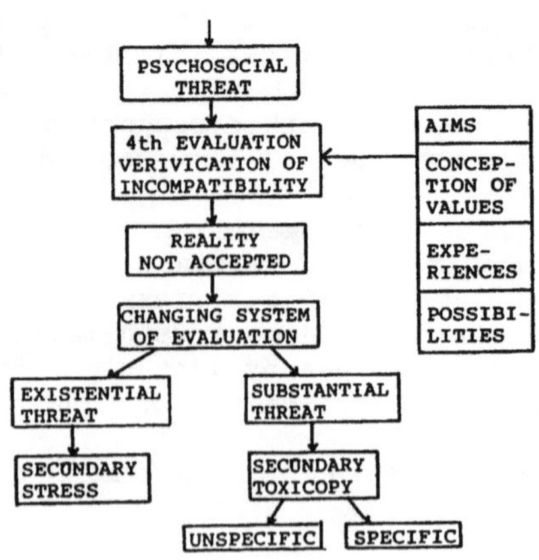

ODOUR CAUSED MASS ILLNESS AROUND A PLANT PRODUCING QUARTZ AND FELDSPAR BY FLOTATION

Lercher, P., Kofler, W.
Department of Social Medicine at the University of Innsbruck, Sonnenburgstraße 16, A-6020 Innsbruck, Austria

INTRODUCTION

In the last ten years mainly US-investigators have reported a series of mass illnesses occuring at different working places. The authors almost unanimously regarded the outbreaks as being of psychic origin as no substantial cause could be found. The odours which often appeared to be the trigger never indicated poisonous concentrations of substances. A similar case seem to be the complaints about the smell of resin near a quartz- and feldspar-processing plant in the federal province of Upper Austria, we will describe below.

THE COURSE OF EVENTS

The only locally relevant emittent in the rural area is a plant standing in the vicinity of industry and settlement since about 1970. A flotation device has been installed in 1975. The odour-stress is caused by an organic mixture resembling the resin of pines which in addition to small amounts of NaF is needed for the flotation. The plant is one of the most important employers in the area.

During the last ten years odour-emissions certainly have risen to a level that is more than three times higher than the original one. According to the law of Weber-Fechner this does not signify a principal changement, whereas the evaluation of the odour-stress in fact has changed principally. It was not until August 1984 that people seriously complained about odour-stress. The complaints were initiated by a neighbour who accused the plant of being guilty of emitting poisons that caused fluorosis of one of his best cows. The necessity of putting down the precious animal drove the farmer to an official accusation. The Federal Government reacted immediately by measuring the emission of the plant, but neighbours doubted the validity of the measurements.

The values were still within the limits defined by VDI. Politicians of different range (community-politicians, members of the Federal Government, members of the State Government) belonging to most different political parties paid the plant several visits. They agreed in the conclusion that there was a serious problem waiting for solution.

As a first reaction one of the authors (Kofler W.) was consulted as medical expert. At this time (February 1985) the plant was especially accused of the odour-stress which lead as far as vomiting of people.

In the meantime public media had taken hold of the problem. The situation was most often presented in a way that let no doubt in the fact that the cattle had died because of fluorosis as a result of emissions of the plant in summer 1985. As further cows fell sick and died in consequence, the plant stopped production. The mass epidemy stopped at once. Intensive official investigation carried out in summer and autumn 1985 did not show any relevant concentrations of poisons. After an evening of information the plant was reopened in January 1986. A new wave of epidemic mass disease rose almost immediately. Police was alarmed in many cases. In August 1986 a device was installed in order to reduce the odorous emissions (efficiency 90 %). Afterwards a reduction of cases could be observed.

MATERIALS AND METHODS

In February 1985 we were not called to carry out a scientific research but to utter our opinion about the level of odour for the mining authority (Berghauptmannschaft). We could only ulilize existing data, documents, objections etc. to describe and evaluate the past situation. This was:

1. The evidences of the local practitioners
2. The expert's reports of the different health officers of the last ten years
3. The results of screening tests carried out in schools and kindergardens during the last ten years
4. The results of medical investigations of the workers of the plant
5. The data of the local police and the local authorities
6. The expert's reports of the juridical proceeding in connection with the death of the cows
7. The results of test programs of children of Linz and its surrounding (3,411 children)
8. The results of different investigations of emission and immission

Further we used the laymen's complaints against the plant which are documented by the local police to describe the second wave of psychogenic mass disease.

RESULTS

We regard it as important to mention the fact that all reports to the police were made in temporal connection with real air pollution by some odorous substance. The evidences (findings) of the local practitioners confirmed the symptoms reported by the individual persons. Medical doctors underlined that the symptoms had not appeared any earlier. They were unable to understand their aetiology. Some of them even reported similiar symptoms themselves. In the last ten years not one case of mottled teeth has been reported even if the caries prophylaxis had not been stopped after the event. Fluoride values in the urine did not differ from non existing population samples. The emission measuring could explain only few percentages of the measured immission levels (table 2 and 3).

But also the immission levels measured in the village were much lower than the health oriented limits. In the surrounding they were higher, probably because of the fluoride emittant 12 kilometers west (the industrial area of Linz) and 8 km east (a large brick producing plant).

Table 3 shows the number of individuals who reported to police according to sex and age. Females expose a slight majority. The number of reports by females is much higher than the one for males (table 4). There is also a great overload of females with regard to the individual frequency of reports (table 5). In table 6 there is listed the prevalence of the different symptoms. It is compared with symptom-prevalences of other epidemic mass diseases. The installation of a biofilter to reduce odorous emission in August 1986 succeeded in decreasing reporting frequency significantly.

DISCUSSION

We have a lot of arguments for the assumption that the information content of the smell rather than toxic emissions of the plant is responsible for the experienced symptoms. The main reasons are:

* The emissions of the plant do not have any measurable effect on immission concentration.
* The immission values are well within the hygiene standards.
* The symptoms disappeared immediately when the plant was shut down.
* When the plant reopened, symptoms reoccured.

Regarding them it appears justified to add the case of St. Georgen to the group of psychogenic mass diseases.

We prefer the signification "toxicopy" as a limitation within the large field of psychogenic mass diseases. Toxicopies, for example, do not necessarily require groups but the mechanisms bringing them about can be effective in every human being. Of course they can easier be recognized and observed in mass occurrances. Additionally, there seem to exist different evaluations of situations. They can initiate psychogenic reactions which in consequence are described as Psychogenic Mass Diseases. The fear of poisons is only one example. Still it appears necessary to make differences between possible evaluations of situations.

Our observations show differences as well as similarities to those of NIOSH which have been made at working places.

Those results are only preliminary ones, as the study is continued. After the installation of the scrubbing device there have been very few complaints. We are paying special attention to those persons who expose symptoms even now.

REFERENCES

(1) Colligan, M. J. (1979). A symposium on the diagnosis and amelioration of mass psychogenic illness in industry. Proceedings of National Institute for Occupational Safety and Health, Cincinnati, Ohio.

(2) Colligan, M. J., Pennebaker, W., Murphy, W. R. (1982). Mass psychogenic illness. A social analysis. Hillsdale, NJ, Erlbaum Publishing Co.

(3) Kofler, W. W.: Toxicopy - a basic mechanism to cope with environmental threats. In: Proceedings of First European Meeting of Environmental Hygiene. Düsseldorf, May 21 - 22, 1987.

TABLE 1: IMMISSION DATA

FLUORIDE CONCENTRATION:		
ST. GEORGEN (NEAR PLANT):	I1:	0,20 µg/Nm³
	I2:	0,50 µg/Nm³
CALCULATED SHARE OF PLANT		≤ 5,00 %
LINZ-STEYREGG (12 KM WESTERN)	I1:	0,76 µg/Nm³
	I2:	2,25 µg/Nm³
MAUTHAUSEN (8 KM EASTERN)	I1:	0,52 µg/Nm³
	I2:	1,50 µg/Nm³
FLUORIDE DUSTFALL:	I1:	369,00 µg/m².d
	I2:	625,00 µg/m².d
CALCULATED SHARE OF PLANT		8,00 µg/m².d
PARTICULATE MATTER:	I1:	0,04 mg/Nm³
	I2:	0,09 mg/Nm³
SO_2: (CALCULATED)	I1:	≤ 0,05 mg/Nm³

P.M., SO_2 (AND NO_x, CO, HC) ACCORDING TO REGULATIONS FOR RECREATIONAL AREAS.

TABLE 2: EMISSION DATA

		QUARTZ	FELDSPAR
P.M.	mg/m³	12	50
FLUORIDE ON P.M.	mg/m³	0,025	0,08
FLUORIDE (GASEOUS)	mg/m³	0,20	0,19
FLUORIDE TOTAL	mg/m³	0,225	0,27
ORG. C.	mg/m³	20	97
NITROSAMINE	µg/m³	N.T.	N.T.
NO	mg/m³	45	25
NO_2	mg/m³	N.T.	23
NO_x	mg/m³	45	48
SO_2 (1%S)	mg/m³	220	220
TEMPERATURE (PROCESS)	°C	89	82
TEMPERATURE (STACK)	°C	51	53
VOLUME	Nm³/H	10.000	10.000

TABLE 3: NUMBER OF INDIVIDUALS, WHO REPORTED TO POLICE CLASSIFIED BY SEX AND AGE

AGE	SEX				SUM	
	MALE		FEMALE			
	N	%	N	%	N	%
11 - 30	3		6		9	13,4
31 - 50	20		22		42	62,7
51 -	7		9		16	23,9
SUM	30	44,8	37	55,2	67	100,0

TABLE 4: NUMBER OF REPORTS TO POLICE CLASSIFIED BY SEX AND AGE

AGE	SEX				SUM	
	MALE		FEMALE			
	N	%	N	%	N	%
11 - 30	3		28		31	8,8
31 - 50	72		155		227	64,5
51 -	27		67		94	26,7
SUM	102	29	250	71	352	100,0

TABLE 5: REPORTS TO POLICE PER INDIVIDUAL CLASSIFIED BY SEX

SEX	REPORTS							SUM
	1x	2x	3x	4x	5x	6x-9x	≥ 10x	
MALE	18	2	4	1	2	0	3	30
FEMALE	8	6	3	1	2	6	11	37
SUM	26	8	7	2	4	6	14	67

TABLE 6: PREVALENCE OF SYMPTOMS REPORTED IN COMPARISON WITH OTHER EPISODES OF MASS PSYCHOGENIC ILLNESS

THIS STUDY		SUMMARY OF 16 NIOSH INVESTIGATIONS *)	NO. REPORTED
HEADACHE	65,7 % (44)	HEADACHE	14
THROAT IRRITATION	65,7 % (44)	DIZZINESS/LIGHT-HEADEDNESS	12
EYE IRRITATION	53,7 (36)	NAUSEA	11
NOSE IRRITATION	52,2 % (35)	DRY MOUTH/THROAT	10
RESPIRATORY IRRITATION	52,2 % (35)	EYE/NOSE/THROAT IRRITATION	9
NAUSEA	37,3 % (25)	SLEEPINESS/DROWSINESS	7
UNSPECIFIC FEELINGS OF THREATENING	37,3 % (25)	WEAKNESS	7
		NUMBNESS/TINGLING	6
MOUTH IRRITATION	19,4 % (13)	CHEST TIGHTNESS/DISCOMFORT	6
SKIN IRRITATION	10,4 % (7)		

GROWTH AND BONE MATURATION OF CHILDREN LIVING IN AREAS WITH DIFFERENT DEGREES OF AIR POLLUTION: A REPEATED STUDY

Pelech L., B. Rosický, H.-W. Schlipköter[1], and R. Dolgner[1]

Institute of Hygiene and Epidemiology, Šrobárova 48, Prague, ČSSR, and Medical Institute of Environmental Hygiene, Auf'm Hennekamp 50, 4000 Düsseldorf, FRG[1]

In a number of surveys conducted during the past twenty years in Czechoslovakia and other European countries, we studied the impact of environmental pollution on the somatic development and the state of health of children living permanently in areas highly burdened with industrial air pollutants.

One of the methods employed in determining developmental effects was the measurement of bone maturation using an X-ray photograph of the children's left hand and wrist. This part of the skeleton was chosen for study because it contains a great number of small bones and hence offers to the investigator detailed and accurate information about a child's biological age. Furthermore, this procedure permits complete protection of other parts of the body, particularly the gonads, from undesired effects of ionizing radiation. Our X-ray apparatus had been specially designed for its use in field studies. Evaluation of X-ray pictures was done according to Greulich and Pyle, but applying standards established for Czech children. Individual findings were expressed as the difference between skeletal and calendary age and means of these differences, in months of acceleration (+) or retardation (-), were calculated for each group. Comparisons were done on the basis of the group means as well as of the relative frequencies of children with bone age retardation of more than 10 months (bone age $< \bar{x} - 1$ SD).

The first studies of this type clearly showed that in children of younger school-age chronically exposed to industrial air pollution, bone maturation was significantly delayed in comparison with children of that age living in relatively clean reference areas; the percentage of subjects with a deficit of more than 10 months, too, differed significantly between groups (table 1, A). Retarded skeletal maturation in school children is taken as the result of the complex influence of a number of environmental factors of which the lack of UV radiation, caused by high airborne dust concentrations, should be one of the most important.

In a collaborative study, the Medical Institute of Environmental Hygiene in Düsseldorf and the Institute of Hygiene and Epidemiology in Prague were able to substantiate this hypothesis. Areas in which surveys had been run in 1974 were again selected for study in 1984 and bone age was determined in elementary school children of the same age range in both rounds. Whereas in the control area skeletal development was found rather unchanged, a marked improvement in this parameter was seen in the industrialized area when comparing the 1974 and 1984 results. It can be read both from the mean values of retardation and the relative frequencies of subjects with a high degree of developmental deficit (table 1, B). During these ten years, atmospheric concentrations of suspended particulates and sulfur dioxide decreased by more than 50 and about 40 %, resp., in the industrialized zone as the result of technical measures and we suggest these changes to be the main cause of the improvement observed in bone maturation.

Table 1

Children's skeletal maturation in terms of a) the mean differences "bone age minus calendary age" in months, b) the percentages of subjects with bone maturation delayed for more than 10 months (bone age < $\overline{x}$ - 1 SD)

A Results from index (polluted) and reference (clean) areas in three European countries

B 1974 and 1984 results from the same index and reference areas in the FRG

A	ČSSR Index	ČSSR Reference	FRG Index	FRG Reference	GDR Index	GDR Reference
Difference	- 6.4	- 1.0	- 7.4	- 1.2	- 10.3	- 2.8
Percentage	32	17	33	13	56	25

B	FRG Index 1974	FRG Index 1984	FRG Reference 1974	FRG Reference 1984
Difference	- 6.5	- 3.9	+ 0.7	- 0.1
Percentage	32	24	12	14

But there remains the question whether the affected individual himself is able to recover when environmental conditions improve. This issue was studied by investigating two simple parameters of erythropoiesis: the concentrations of red blood cells and hemoglobin. Once having established the detrimental effect of environmental pollution on these functional indicators in school children (table 2, A), we looked for the impact a temporary stay in clean air would have on them. Czech school children residing in problem areas regularly are sent to rural districts where they continue lessons for three weeks on the average. In evaluating the health benefit resulting from this compensational measure, we found good evidence of functional improvement when comparing mean values as well as percentages observed shortly before and shortly after country school time (table 2, B).

Table 2

Red blood cell and hemoglobin concentrations in children in terms of a) the mean concentrations ($x\ 10^{12}$/l and g/dl, resp.), b) the percentages of subjects with values $< \bar{x} - 1$ SD

A Results from an index (polluted) and a reference (clean) area

B Results of determinations before and after a three weeks stay at a non-polluted country school site

	RBC		Hb	
A	Index	Reference	Index	Reference
Concentration	4.48	4.71	11.4	12.3
Percentage	43	18	44	13
	RBC		**Hb**	
B	Before	After	Before	After
Concentration	4.62	4.87	12.3	12.8
Percentage	36	14	42	15

On the basis of these results and mainly by considering the differences seen between groups in the relative numbers of children more strongly affected by environmental noxae, we arrived at the problem of the so-called risk groups, their determinants, and the ways open for their protection. Intensity of air pollution held constant, this subgroup of children at special risk is the greater, the more susceptible the human organism, the more restricted the possibilities for its recovery, and the greater the overall dose of noxious substances absorbed. It comprises, therefore, the younger, those otherwise exposed, the socially affected (less family care), the genetically predisposed (inborn somatic, metabolic, immunologic defects), and those in bad health (pre-existing chronic illness or frequently recurring acute illness) and its menbers are at higher risk for developing adverse health effects than children without the above mentioned "negative traits", when exposed to the same concentrations of air pollutants.

Knowledge of this causational web and experience gained with the compensational measures introduced up to now clearly indicate the necessity for devising a complex system of care for children living in polluted areas. The two basic components of this system, i.e. family care and social measures, must form an entity in which the ratio of one part to the other will vary in dependance on the kind and degree of environmental pollution, and the child's personal and family characteristics.

REFERENCES

(1) Dolgner, R., L. Pelech, H.-W. Schlipköter, P. Schmidt (1971) Studium vlivu znečištění ovzduší na zdravotní stav dětí v Porúří, Čs.Hyg., 16/2-3, 62-70.

(2) Pelech, L. (1970) Verzögertes Knochenalter bei Stadt- und Landkindern in Gebieten mit vermehrter Luftverunreinigung, Fortschritte der Medizin, 88/12, 510-516.

(3) Pelech, L. (1985) Die Entwicklung von Kindern in luftverunreinigten Gebieten und eine Möglichkeit ihrer Beeinflussung, Sozialpädiatrie in Praxis u. Klinik, 7/3, 128-131.

(4) Schmidt, P., R. Dolgner (1977) Zur Deutung einiger Befunde aus Kinderuntersuchungen in Gebieten mit unterschiedlich starker Luftverunreinigung, Zbl. Bakt. Hyg. I. Abt. Orig. B, 165, 539-547.

(5) Thielebeule, U., L. Pelech, P. Grosser, K. Horn (1980) Körperhöhe und Knochenalter bei Schulkindern in lufthygienisch unterschiedlich belasteten Gebieten, Z. ges. Hyg. u. Grenzgeb., 26/10, 771-774.

(6) Volkotrub, L.P. (1979) Vlijanije atmosfernych zagriazněnij na sostojanije periferičeskoj krovi u detej, Gig. i San., -/9, 73-74.

CHROMOSOME ABERRATIONS IN PERIPHERAL LYMPHOCYTES OF PERSONS PRESENT IN THE VICINITY OF CHERNOBYL DURING AND AFTER THE REACTOR ACCIDENT

Stephan, G., U. Oestreicher

Institute for Radiation Hygiene of the Federal Health Office

Ingolstädter Landstr. 1, D-8042 Neuherberg, FRG

INTRODUCTION

Chromosome analysis as a supplement to physical dosimetry to-day is an important radiation protection tool for the identification of radiation exposures. The effectiveness of this method, also in the low dose range, is substantiated by literature reports (1,2,3) as well as by own chromosome examinations. Our results have shown to be in good agreement with those from physically determined whole body doses (in preparation).

In consequence of the reactor accident in Chernobyl, 87 blood samples were received of persons who had been physically present in the Chernobyl vicinity, mainly on account of industrial work assignments by West-German contracting firms. Although based on about 1500 evaluated cells, the lower detection limit for low LET-radiation is approximately 0.07 Gy and physically evaluated doses were well below this limit, the analysis was made mainly on account of the interest the firm had expressed in this type of investigation.

MATERIALS AND METHODS

The methods used for cell culture, sample preparation and chromosome analysis have been reported earlier (4).

RESULTS

Of the various structural types of chromosome aberrations, the dicentric chromosome (dic) is best suited for the detection of radiation exposure and, therefore, only this aberration type is given.

To asses the spontaneous frequency of dicentric chromosomes in peripheral lymphocytes, 13 897 cells have presently been analysed in our laboratory and 13 dics evaluated. Accordingly, the near mean frequency is 1 dic/1000 cells. The examination included 23 persons ranging in age from 0-59 years.

In the following we limited ourselves to results from the chromosome analyses of 27 persons who had returned from Mogilev (300 km north of Chernobyl) after the reactor accident, mainly because a large group of persons coming from the same area had been examined here. At least 600 cells per person were analysed. The number of dics found in the individual persons is between 0 and 9. On the overall we evaluated 26 080 cells and analysed 85 dicentric chromosomes. The resulting mean aberration rate of 3.26 dic/1000 cells is significantly different from the spontaneous rate ($p = 1.4 \cdot 10^{-6}$).

Figure 1 shows a comparison of both groups (control-group, Mogilev-group) with a set aberration rate. For the Mogilev-group, only the first 300-400 evaluated cells were taken into consideration because this number of cells per person was scored in the control group. A lower percentage of persons with 0 dic is clearly shown by the Mogilev group compared to the control group.

SUMMARY AND CONCLUSIONS

For the purpose of drawing a relationship between the results and the reactor accident in Chernobyl, we had to establish whether the examined persons were subjected to medical or occupational radiation exposures prior to the Chernobyl accident, or were known to have taken pharmaceutical products. These data were obtained by use of a questionnaire.

In reference to the reported pharmaceuticals (Ulcogant, Aspirin, migraine-tablets), it may be assumed that these did not contribute to an increase in the aberration rate. The medical radiation exposures, too, should not have resulted in an increased spontaneous rate, since the number of dicentric chromosomes is within the range of the spontaneous frequency for persons with comparatively frequent diagnostic x-ray exposures. One person should be excluded because in this case, dicentric chromosomes may also have been induced by an x-ray examination of the kidney using contrast medium (5).

Occupational, exposures had occurred with 2 of the examined persons, whereby 1 individual had received an officially substantiated dose of 3760 mrem. Another person may have been occupationally exposed to radiation - the exact clarification of data is still pending.

The exclusion of the above described 4 individuals on account of the uncertainty in assigning a set number of dicentric chromosomes to the x-ray examinations or the occupational exposures, results in a mean aberration rate of 2.57 dic/1000 cells (22 173 analysed cells, 57 dicentric chromosomes). This rate is significantly different from the spontaneous rate ($p = 2.7 \cdot 10^{-4}$). Based on our calibration curve for 250 kV x-rays $y = ((5.0 \pm 0.11) \cdot 10^{-2}\ D + (7.7 \pm 0.09) \cdot 10^{-2}\ ) \cdot D^2\ ))$ an equivalent whole body dose of 0.05 Gy may be assumed from this rate.

This dose may have been influenced by incomplete information given by the questioned persons. For example, medical or occupational radiation exposures important for the interpretation were not reported. In that case the above dose value would be overestimated. In total, the Mogilev-group shows a clear deviation in the number of dics per person in favor of classes with more dics, as illustrated in figure 1.

This finding emphasizes the fact that persons who returned from Mogilev were likely to be exposed with doses higher than those physically estimated of less than 0.5 mSv (6).

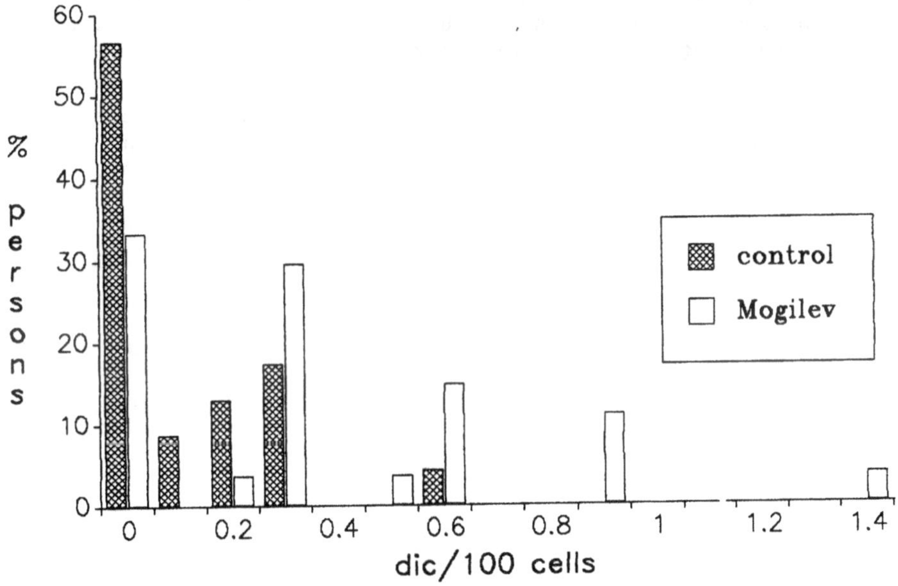

FIG.1: Persons with identical aberration rates

REFERENCES

(1) Evans, H.J., K.E.Buckton, G.E.Hamilton and A. Carothers (1979) Radiation-induced chromosome aberrations in nucleardockyard workers, Nature, 277, 531-534.

(2) Lloyd, D.C., R.J. Purrott, E.J. Reeder (1980) The incidence of unstable chromosome aberrations in peripheral blood lymp hocytes from unirradiated and occupationally exposed people (1980) Mutat.Res., 72, 523-532.

(3) Bauchinger, M., J. Kolin-Gerresheim, E. Schmid, J. Dresp Chromosome analyses of nuclear-power plant workers (1980), Int.J.Radiat.Biol., 38, No. 5, 577-581.

(4) Stephan, G., I.D. Adler, D. Schwartz-Porsche, K.-U. Hollihn, G. Obe (1979) Charaterization of feline whole-blood cultures and determination of the frequency of radiation-induced dicentrics in human and feline lymphocytes, Int.J.Radiat.Biol., 35, 351-359.

(5) Hadnagy, W., G. Stephan, F. Kossel (1982) Enhanced yield of chromosomal aberrations in human peripheral lymphocytes in vitro using contrast media in x-irradiation, Mutat. Res.,104, 249-254.

(6) USSR State Committee on the Utilization of Atomic Energy "The accident at the Chernobyl Nuclear Power Plant and its Consequences", International Atomic Energy Agency (post accident review meeting), Wien, 25. - 29.8.1986

DISPOSITIONS PROPHYLAXIS FOR CHILDREN FROM AIR POLLUTED AREAS

Thielebeule, U. and Christel Hülße
Institut of Hygiene, County of Rostock,
Gertrudenstraße 9, 2500 Rostock, GDR

INTRODUCTION

Investigations with groups of children which have been carried out in the GDR for two decades have shown - like in other countries - that long term exposure to air pollutants results in a variety of non specific effects on children / 4, 5 /.
By comparison with children living in non polluted surroundings we found alterations of haematological parameters (erythrocytes, leucocytes, haemoglobin and haematocrit) / 1 /.
The same was true for immunological criteria as rate of phagocytosis of polymorphonuclear neutrophile leucocytes, phases of activity of the nucleoli of peripheral lymphocytes, titer of tetanus-antitoxin, immunoglobulins, complement C_{3c} and C_4, serum-lysozyme-titer, α_1-antitrypsin and transferrin / 1 /.
The clinical examination and testing of the function of the lungs also revealed differences between exposed and nonexposed groups of children / 5 /.
Testing several psychological procedures we found the concentration test according to KURTH / 3 / most suitable / 2 /. With this method deviations of the concentrativeness as well as of the quality of performance could be demonstrated in children originating from highly loaded regions.
But all the deviations found were still within the range of the very broadly accepted "Norm" for all groups of children and could thus not be taken as clear prepathological signs but only as mechanism's of defence or adaption of the organism.
Measures of dispositionsprophylaxis for children from polluted area are aimed at increasing levels of reristance and the stabilisation of health.

MATERIALS AND METHODS

In order to be able to consider these alterations as caused by exposure to air pollution, we moved whole classes of schoolchildren (at six to eight years of age) in February/March for four weeks from polluted areas to non polluted surroundings at optimum conditions, including extended stays at open air (cure groups). At the same time comparable groups in the polluted area and control groups in clean air regions were examined by the same research team and the same methods (Tab. 1).

Tab. 1 Methods of investigation

1. Test for the examination of the concentrativeness to KURTH / 3 / .
2. Lung-Function (FVC, FEV_1)
3. Haematological parameters: erythrocytes, leucocytes, haemoglobin, haematocrit, nucleoli-activity of peripheral lymphocytes
4. Immunological parameters:
   rate of phagocytosis of polymorphonuclear neutrophile leucocytes, immunoglobulins A, G and M, complement C_{3c} and C_4, transferrin, α_1-antitrypsin, serum-lysozyme, T-lymphocytes, T-helper- and T-suppressor-cells.

RESULTS

Results demonstrate improvement of the concentrativeness of children from industrial areas after four weeks clean air stay. Investigation of the lung function showed a significant increase of the FVC- and FEV_1-values.
In comparion with the clean-air situation significantly decreased values of the red blood picture (erythrocytes, haemoglobin, haematocrit) return to the norm at the end of the cure. This effect could be demonstrated to be lasting up to half a year after the stay in clean air (Fig. 1).

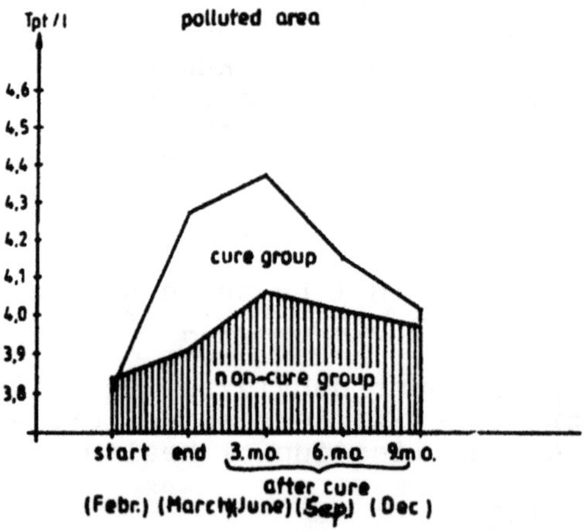

Fig. 1 Erythrocytes - value of children from industrial areas - cure- and non-cure-group -

The evaluation of morphological form of the nucleoli of the peripheral lymphocytes at the end of the clean air stay resulted in findings approaching those obtained with children originally living in non polluted surroundings and lasted up to 9 months after the termination of the clean air stay (Fig. 2).

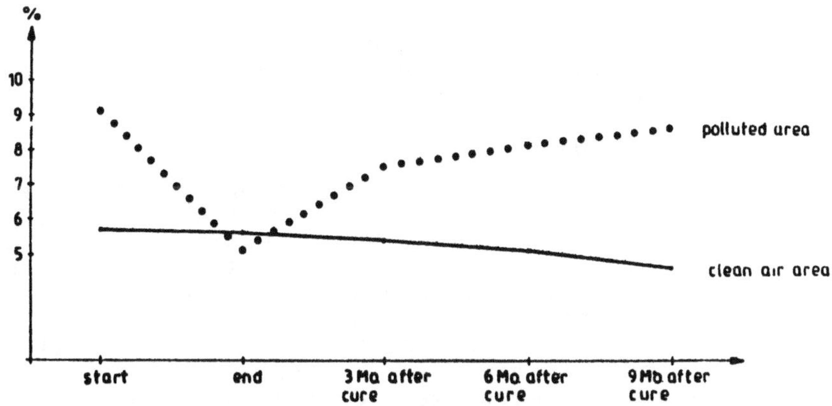

Fig. 2 Lymphocytes with RNS-active nucleoli from exposed children before and after cure in comparison with control group from clean air area

Similar alterations were also found for the immunoglobulins responsible for the non specific humoral immunity, for the serum proteins α 1-antitrypsin and transferrin (Fig. 3), but not for complement-titer.

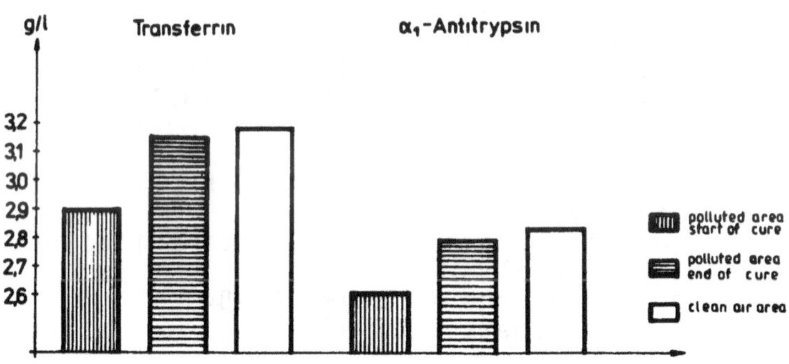

Fig. 3 Transferrin - and α 1-antitrypsin-values of children from polluted areas before and after cure in comparison to children from non polluted area

Increase of T-helper-cells in children from polluted areas after cure-end shows the influence of clean air stay on the specific cellulare immunity.

Therefore the improvement of the conditions approaching those prevailing in regions which we consider as "norm" areas which does not mean the absence of any pollution of course are regarded as normalisation of the organism's response.

Approximately the same alterations could be observed in polluted surroundings during the summer vacations of the children. This effect can be reached some months earlier by properly planned stays in clean air areas.

SUMMARY AND CONCLUSIONS

The deviating conditions in children coming from heavily loaded industrial areas - although being within the physiological norm - must be regarded as indication for the stress of the organism and its efforts for adaptation. Therefore repeated application of clean air stays can have cinditioning impacts on the organism and can possibly avoid the transition from a reaction of adaptation to premorbidity. In spite of all efforts aimed at the protection of the environment effective improvement of the conditions of exposure can only be expected within decades. Based on this fact we are now trying to find out by further studies the optimum approach for duration and intervals of such methods of disposition prophylaxis and to develop them in to real preventive measures.

REFERENCES

(1) Hülße, C., D. Nehls and U. Thielebeule (1984) Haematologic and immunologic examinations during a cure in children from an industrial concentrated region with hygienically loaded air, Z. gesamte Hyg.30, 710 - 712

(2) Hülße, C., U. Thielebeule and E. Kurth (1987) Epidemiologic investigations to the influence of air pollutions on concentrativeness of school children, Z. gesamte Hyg. 30, 286 - 288

(3) Kurth, E. (1985) A test series for the examination of the concentrativeness. Psychodiagnostic centre of the Department of psychology Humboldt-Univ., Berlin

(4) Schlipköter, H.-W. and K. Beyen (1984) Methods of investigation of effect of air pollution on the people, Gesundheitsingenieur, 105, 175 - 186

(5) Thielebeule, U., C. Hülße, D. Wolf, S. Schulz and A.Steinhauser (1985) Preventive measures for children resident in industrial centres, Z. gesamte Hyg. 31, 174 - 176

V. ENVIRONMENTAL CONTROL AND LEGISLATION

SOURCE APPORTIONMENT OF HEAVY METALS IN AIR PARTICULATE MATTER USING AUTOMATED ELECTRON PROBE MICRO ANALYSIS.

Van Borm W. and Adams F.

Department of Chemistry, University of Antwerp (UIA), Universiteitsplein 1, B-2610 Wilrijk, Belgium.

INTRODUCTION.

Source apportionment of air particulate matter, using receptor models recommended by the Environmental Protection Agency, USA, are not well adapted for situations with a complex source composition because of the inherent inability to resolve chemically equivalent sources. Alternatively, methods allowing the analysis of single particles such as electron probe micro analysis (EPMA) provide direct information on chemical composition and morphology for each individual particle and thus, in the most optimal situations for each individual source. It has therefore been suggested to use these methods to provide information in addition to that obtained in the analysis of bulk samples[1].

As part of a research project on heavy metal air pollution in Northwest Limburg, Belgium, a method was developed to assess the impact of several heavy metal sources, located within one zinc smelter complex. As was shown from deposition studies, this smelter is responsible for elevated deposition rates of Pb, Zn and Cd, the latter two estimated at 32 t/y and 320 kg/y respectively[2]. Attention was focused on the possibility to discriminate between smelter-related and non-related heavy metal dust to assess on the one hand the current impact of the smelter on the environment and on the other hand the so called 'historical' pollution, namely the impact of resuspended soil dust.

EXPERIMENTAL.

Sampling and Analysis.

In February 1986, 10 air particulate samples were taken at 3 collocated sites near the zinc smelter, in the direction of the prevailing wind. Particulate matter was collected for about half an hour on Nuclepore polycarbonate membrane filters (25 mm x 0.4 µm) mounted in an open filter head at a filter flow rate of 2.5 lpm. They were mounted on plexiglass stubs and were carbon-coated before analysis.

Electron probe micro analysis was performed on a JEOL JXCA-733 Superprobe (Jeol Ltd, Tokyo, Japan) equipped with an energy dispersive X-ray detector. Automated analysis was performed using a software program, developed at the University of Antwerp. A full description of the analytical procedure was given by Raeymaekers et al.[3]. Elemental concentrations and particle densities were calculated by a fast standardless ZAF correction method, providing quantitative analytical information[4].

Data Processing.

The large data sets, composed for each particle of 16 elemental concentrations and the projected average diameter were interpreted using a combination of both hierarchical and nonhierarchical cluster analysis. This method groups particles of similar composition into clusters without relying on a priori chemical criteria. Nor does the method use any knowledge of predefined particle classes. A detailed description of this method was published by Van Borm et al.[5].

RESULTS AND DISCUSSION.

In view of the adverse health effects of suspended particulate matter following the intake by and the deposition in the human respiratory system, we decided to split up the datasets artificially into two subsets, one containing particles with an aerodynamic diameter below 2.5 µm, the other containing particles above 2.5 µm. The fine particles are mainly deposited in the alveolar region and to a lesser extent in the tracheobronchial system, while the coarse particles are deposited in the head region. The aerodynamic diameters were calculated from the projected diameter and the particle density.

For the fine particles 10 clusters were obtained that could be representative for some of the particle sources (Table 1). We observed particles with an Al-Si-Fe composition, typical for soil dust, zinc-contaminated soil particles with high concentrations of Zn, particles with a distinct Zn-S composition, representative for the ZnS ore that is being processed at the smelter, Fe-particles from the goethite residue which is a waste product dumped on the factory premisses, mostly containing Fe-oxides. Pb-Zn-Fe particles along with some other non-heavy metal particles from different sources are also present. The mean diameters ranged from 1.8 µm to 2.0 µm.

Table 1 : Composition (weight%) and abundance (% of particles) of the clusters for the fine mode particles.

#	Abundance	Source	Composition
1	28.6 ± 4.8	Soil	Al(25),Si(55),Fe(9)
2	19.7 ± 2.9	Zn-contam.soil	Al(10),Si(20),Fe(29),Zn(20)
3	11.6 ± 3.1	Goethite	Fe(69),Zn(10)
4	11.1 ± 5.4	Pb-Zn-Fe-part.	Si(7),Pb(49),Zn(6),Fe(14)
5	7.2 ± 1.1	ZnS-ore	Zn(74), S(13)
	21.8 ± 9.9	Others	

For the coarse mode particles similar results were obtained, excepted that the Pb-Zn-Fe particles the most abundant particle category. Also, some other particle types were detected such as pure Pb-particles (Table 2). Mean diameters ranged from 9 to 12 µm.

The fraction of heavy metal containing coarse particles was definitely higher in the fine particle fraction which indicates a higher nuisance for the nasal pharyngal system.

Table 2 : Composition (weight%) and abundance (% of particles) of the clusters for the coarse mode particles.

#	Abundance	Source	Composition
1	30.9 ± 6.0	Soil	Al(21),Si(53),Fe(14)
2	20.3 ± 4.4	Pb-Zn-Fe-part.	Si(12),Pb(33),Zn(12),Fe(25)
3	15.2 ± 1.8	Goethite	Fe(63),Zn(14)
4	14.4 ± 2.2	Zn-contam.soil	Al(10),Si(21),Fe(26),Zn(27)
5	7.0 ± 7.9	ZnS-ore	Zn(74), S(11)
6	6.7 ± 3.4	Pb-particles	Pb(69)
	5.3 ± 1.6	Others	

A comparison of the distribution of the clusters on the locations revealed that the overall abundance of zinc ore particles is about 2 %, while abundances as high as 15 % were measured at 100 m downwind from the storage yard, but decreasing with distance from the source. For the Pb-Zn-Fe-source a mean abundance of 17 % was measured and was found to be highest (24 %) along the main road passing in front of the smelter. It is therefore likely that these latter particles were originating from resuspended road dust.

Zinc and Lead Pollution.

When all particle clusters were divided into two catagories respectively "non related" and "related" to the smelter (Table 3), according to the criterium that clusters that can not be pinpointed to known zinc sources in the smelter belong to the "non-related" category, it became clear that the zinc smelter contributes for about 20 % to the TSP for both fractions. We can call the other category the 'historical' pollution, because it is composed of the zinc-contaminated soil particles. It contributes for about 33 % to the particle content of the inorganic fraction of the TSP. The mean zinc concentration is 12.5 %, which is very high when compared with zinc levels in residential areas that are about 1 %.

The same analysis was made for the lead containing particles. Thus it was show that about 19 % of the particles contain lead, most of these being coarse particles. The mean lead concentration is 8 %, which is also very high when compared to background levels.

Table 3 : Concentrations (%) of zinc and lead and abundancies (% of particles) of zinc- and lead-containing particles in the inorganic fraction of the TSP.

	Zn	Zn-part. related	Zn-part. non-related	Pb	Pb-part.
fine	10	18.8	30.8	5	11.1
coarse	15	22.2	34.7	11	27.0
mean	12.5	20.5	32.8	8	19.1

CONCLUSION.

More than 4,500 particles were analyzed and grouped into clusters, each representing a particulate source. Heavy metals such as lead and zinc were found to be either associated with suspended soil dust or else in their original form as emitted through the sources, as was the case with the Zns ore particles. Despite the presence of an operating cadmium refinery no cadmium containing particles were detected.

REFERENCES.

1. Gordon G.E., Pierson W.E., Daisey J.M., Lioy P.J., Cooper J.A., Watson J.G.Jr., Cass G.R. (1984), Considerations for design of source apportionment studies. Atmospheric Environment, 18:1567-1582.

2. Demuynck M., Verstraeten J. (1985), Areawide distribution of atmospheric heavy metal deposition in Northwest-Limburg. "Proceedings SCOPE-symposium : Belgian research on metal cycling in the environment.", Brussels, p.47-57.

3. Raeymaekers B. (1986), Ph.D-Thesis, University of Antwerp (UIA), Belgium.

4. Raeymaekers B., Van Espen P., Adams F. (1987), A fast standardless ZAF correction for electron probe micro analysis. Anal.Chem., in press.

5. Van Borm W., Adams F. (1987), Cluster analysis of elec tron microprobe data of individual particles for source apportionment of air particulate matter. Environ. Sci. Technol., submitted.

New Aspects to Define Air Quality Guidelines for Carcinogenic Substances in the Federal Republic of Germany

Dobbertin, S.

Umweltbundesamt, Bismarckplatz 1, D-1000 Berlin 33, FRG

Preliminary remark: it is true, I never was good at math in school. Therefore, please do not expect me to elaborate on mathematical formula for risk assessments. The purpose of my lecture is merely to present different ways of determining Air Quality Guidelines. Thank you for your cooperation.

In the Federal Republic of Germany, 25.000 people die from lung cancer annually; these figures are continuing to rise. Especially men are affected. Smoking is considered the main cause for this deathrate, and, along with exposure to carcinogenic substances at the work place, air pollution is another significant cause of lung cancer.

A quantification of immissions-related cancers, however, poses problems because of methodological shortcomings. Regardless of whether 2.000, 5.000 or 10.000 people die annually from lung cancer caused by air pollution, the severity of this disease requires its causal factors to be recognized, controlled and eliminated. An initial step towards this end was taken by the German Federal Government in 1983 and 1986 by imposing limitations on the emissions of a few carcinogenic substances from installations subject to licensing. However, this limitation has not yet produced a satisfactory minimization of the cancer risk, as it does not take contributing factors, such as prior exposure and other emission sources, into account. Thus, for the last ten years, the requirement was raised to set threshold values also for immissions of carcinogenic air pollutants, which would reduce the risk more effectively. In 1978, for example, the Medical Institute for Environmental Hygiene proposed an immission limit value for arsenic (25 ng/m^3) and for benzo(a)pyrene (10 ng/m^3).

The Federal Environmental Agency (Umweltbundesamt) supported these requirements in its Air Quality Criteria documents. A document on benzene was published by the Umweltbundesamt in 1982, recommending an immission limit value of 30 $\mu g/m^3$. These proposals were based on prior exposure and on economic feasibility. Yet, the Federal Government has been hesitant to establish immission limits for carcinogenic air pollutants, the reason being that it has so far not been possible to establish a biological-toxicological safety threshold for carcinogens. The establishment of immission standards would contradict the provisions of the Federal Air Quality Control Act, the purpose of which is protect man's health. This, however, could only be ensured by setting concentrations below the hazardous level. The World Health Organization, in its Air Quality Guidelines, argues along the same lines: "If a substance is a proven human carcinogen, no safe level can be proposed because a threshold is not known to exist. Therefore, exposures should be kept as low as possible." The WHO, however, does not leave it at this global demand, but offers lifetime risks to facilitate decision-making on the establishment of guidelines.

A new course for determining immission standards is being followed in the Federal Republic of Germany. It starts out from a cancer risk considered tolerable for the population at large, which is arbitrarily set at 1 : 1000 as lifetime risk for all carcinogenic air pollutants. On the basis of available measurements of immissions, seven carcinogenic air pollutants were determined: arsenic, asbestos, cadmium, benzene, benzo(a)pyrene, chromium, and nickel. With respect to the aggregate effects of these substances, an additive effect mechanism is assumed.

With the aid of unit risks, the concentration in air for each of these carcinogens is calculated in a manner that precludes exceeding the risk of 1 : 1000. Let me illustrate this with benzo(a)pyrene as an example:

Assuming an even distribution of the total risk among the seven substances, the individual risk can be calculated ($1.43 \cdot 10^{-4}$). This individual risk together with the unit risk ($7 \cdot 10^{-2}$ per 1 $\mu g/m^3$ after inhalation ≙ 0.14 ng/m^3 per cancer case in 100.000 persons exposed) could then be used directly to calculate the immission load which may not be exceeded in order to remain within the risk. It would, however, be too simple to proceed as described, because such a calculation does not take prior exposure and certain imponderables into account. Imponderables would, for instance, be the occurrence of new carcinogenic substances and possible over-additive effects of mixtures of substances. In a differentiated derivation, the imponderables are taken into account in as much as it provides for only 70 % of the risk, leaving a safety margin for only 70 % of the risk. In this calculation, prior exposure is considered in the form of the highest annual mean value.

Thus, the differentiated derivation of an immission limit value for a carcinogenic air pollutant proceeds as follows: From prior exposure and unit risk, its percentage in the total risk is determined and the safety margin considered; this is then put in relation to the unit risk in order to calculate the immission limit value.

In conrete terms, this means for benzo(a)pyrene: Cancer risk is 1 person in 100.000 at 0,14 ng/m³ that is the unit risk, but not related to 1 μg/m³ but to 1 person in 100.000. Prior exposure is 10 ng/m³.
The total hazard potential is the quotient

$$\frac{\text{prior exposure}}{\text{cancer risk}}$$

That is 71,4 %. Because the result of the quotients of the 7 carcinogenic substances is altogether 154 %, we have to reduce it on 100 %. That means that the total hazard potential for BaP is really 52,5 %. Now we have to take into account the safety margin of 70 %. Than we have to put 52,5 % of the safety margin in relation to the cancer risk and get the immission limit value of 5 ng BaP/m³.

I hope that derivations of this type make it possible for politicians to act. And if, in addition to all that, you decide to stop smoking, then we will have solved the problem of bronchial carcinoma for the most part.

PRINCIPLES AND METHODS FOR CONTROL OF AIR POLLUTION IN HEALTH RESORTS

Römmelt,H.[a], K.Dirnagl and A.Schuh[b]
a)Institut u.Poliklinik f.Arbeitsmedizin b)Inst.f.Med.Balneologie u. Klimatologie der Universität, Marchioninistr.17, D-8000 München 70

INTRODUCTION

An essential part of the beneficial effects of cures in health resorts can be attributed to the transfer of the patient to an environment promoting recovery to good health. And among the environmental components the air quality is certainly an important one.

The procedure for qualification of a place as a health resort included regulations to ensure freedom from air contaminants as early as in 1963 (1). An attempt had been made to define the permissible pollution at 20 or 40% of the concentration limits for gaseous pollutants set by the German "TA-Luft". In practice, however, these regulations have never been executed, mainly because neither the financial means nor the manpower and equipment for measurement were available. The only expedient for getting some control over the air quality had been a primitive system of dust sampling and evaluation.

MEASURING METHOD

New developments in passive sampling methods of gaseous pollutants have changed the situation. For instance, the technique of Surface Activated Monitoring (SAM) adopted by the "Umweltbundesamt" (4) promised to satisfy many of the requirements for practicable air quality control in health resorts. Some of these special requirements are:

* Very low detection limits * Independence from electric mains supply * Minimal demands on attendance for the sampling devices * Bearable costs of equipment and analytic procedures even for long periods of monitoring and many sampling locations.

However, the before mentioned technique , using glass fiber filters soaked with potassium carbonate as absorbent for airborne gases, is limited in its usefulness for evaluation of air quality in health resorts by the following properties: * A sampling period of 4 weeks, reducing the possibility of analyzing the influence of weather conditions and of transitory emission events * A strong dependance on wind velocity of the gas deposition rate on the probe which had not yet been quantitatively examined. * Lack of experience about the correlation between concentration in the air and deposition on the SAM-probes for other substances than sulfur dioxide. Apart from heating, motor vehicles are the prevailing source of air pollution in health resorts. That is why there is also a strong need for some means of measuring nitrogen oxides.

The results of a 3 years effort to overcome as much as possible the limitations can be shortly summarized as follows (3): 1) As a consequence of refinements of the analytical procedure, adequate precision of the results can be obtained even in a locally uncontaminated environment with sampling periods of one week. 2) Oxidative pretreatment of exposed samples before the ion chromatographic analysis overcomes some difficulties originating from the fact, that nitrogen dioxide is caught partly as nitrite, partly as nitrate. 3) The dependance of the sampling rate on ventilation of the probe has been determined in the laboratory. When applied to measurements in the field, an approximate correction for the local ventilation can be applied.

Comparative tests at several continuous monitoring stations were performed for at least one year with these modifications . The air quality at the stations covered the range from almost background contamination to the situation amidst a large town. The conformity between weekly pairs of SAM and concentration data proved to be sufficient for correct reproduction of the seasonal trends as well as of weather dependent episodes of changing air pollution. The correlation coefficient between yearly means amounts to 0.94 for SO_2 as well as for NO_2.

Fig.1 gives an example of the information obtained from a series of measurements at 3 sites of a health resort. The main immission peaks are not produced locally, they are caused by transportation of contaminants from distant sources.

Fig.1 NO_2 Immission Rates at 3 Sites of a Health Resort derived from 1-week Expositions of SAM-Probes. 1 = Traffic Center, 2 = Residence Area, 3 = Area for Open Air Therapy

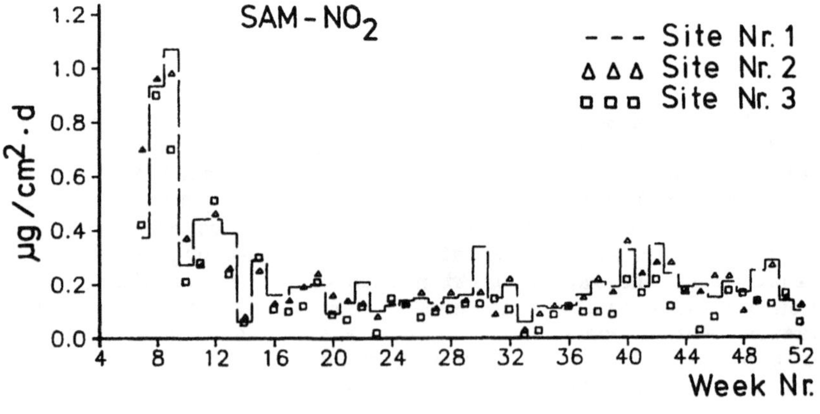

In order to further improve the method toward less dependance on ventilation, we are just testing a diffusion controlled modification of sampling.

EVALUTION OF AIR QUALITY

Having obtained a means for estimating the air pollution level at several sites of a health resort, the next question is how to differentiate between admission or rejection of a claim for qualification.

To that end a classification score on the basis of the 1-year mean and the 95 percentile of a measuring series at representative sites of the place has been established between two bonds: A lower one - marked by a score of zero is equivalent to the concentration level at background stations - and an upper one with a score of 4 is equivalent to the limits of the widely accepted guideline "VDI 2310", extrapolated to a sampling period of one week. The following tables shall give an insight to the system of classification:

Table 1: Scores of Air Quality and yearly Concentration Averages

	SO_2	NO_2	
0 = virtually pure air			
1 = hardly perceptible contamination	up to 15	up to 12	ug/m³
2 = low contamination	" " 25	" " 20	"
3 = moderate contamination	" " 50	" " 40	"
4 = population tolerance limit	" " 100	" " 80	"
5 = very high contamination	exceeding 100	exc. 80	"

The rules (2) for approval of an appropriate air quality are shown in condensed form in the next table:

Table 2: Requirements of Air Quality in different Areas of Health Resorts

	Maximum Scores for		
	Traffic Center	Residence Center	Open Air
Thermal Stations	4	3	2
Thermal Stations for Respiratory Diseases	4	2	2
Climatic Stations	3	2	1

The requirements are graduated according to the therapeutic objective of a station. The most exacting conditions are imposed to the surroundings of climatic stations, but even watering places should always have at least one area of particularly good air quality. Within the next years, all existing health resorts in Germany will be checked for air quality according to the measuring and classification system described.

SUMMARY

Early attempts to set air quality standards in German health resorts had suffered from a lack of practicable methods for longtime measurement of air contaminants in several areas of each station. Therefore, a passive sampling technique (SAM = Surface Activated Monitoring) has now been adapted to the special requirements. SO_2 and NO_2 immission rates are determined by series of 1-week exposures covering at least one year.

The results allow a classification of air quality by 5 scores in the range between background contamination and the limits of guideline VDI 2310. They are forming the basis of new regulations of the "Deutscher Bäderverband". Permissible maximum scores are differentiated according to the therapeutic objective of a station and to the importance of the area of measurement for the patients. A short outline of the regulations is given. All German health resorts are obliged to have checked their air quality using this system of measurement and evaluation.

ACKNOWLEDGMENTS

The following institutions have supported our work:
Deutscher Bäderverband e.V., Bonn.
Bayerisches Staatsministerium für Landesentwicklung und Umweltfragen, München.

REFERENCES

(1) Deutscher Bäderverband e.V. (1963) Richtlinien und Begriffsbestimmungen für Kurorte, Erholungsorte und Heilbrunnen. (Internal regulations, with revisions in 1972 , 1979 and 1987)

(2) Deutscher Bäderverband e.V. (1987) Begriffsbestimmungen für Kurorte, Erholungsorte und Heilbrunnen: Neufassung des Abschnitts 33 "Bioklima und Luftqualität" mit Anhang zu den Ziffern 3314 und 332, Heilbad und Kurort 9, 182 - 186

(3) RÖMMELT,H., K.DIRNAGL and A.SCHUH (1987) Luftqualität in Heilklimatischen Kurorten - Meßverfahren - , Wiss. Mitteilungen des Meteorologischen Instituts der Universität München: Tagungsbericht "2.Treffen des Arbeitskreises Humanbiometeorologie (in print)

(4) RUMPEL, R., (1984) Ein Verfahren zur Feststellung von Immissionsraten: Oberflächenaktive Monitore (SAM), Monatsberichte aus dem Meßnetz H7/84, Umwelt-Bundesamt Berlin

ON THE EMISSION OF FIBROUS PARTICLES FROM CORRODED ASBESTOS-CEMENT PRODUCTS

Spurny, K.R.
Dept. of Aerosol Chemistry
Fraunhofer Institute for Environmental Chemistry and Ecotoxicology,
5948 Schmallenberg-Grafschaft, FRG

INTRODUCTION

Asbestos cement products, e.g. roof tiles, contain as much as 11 to 15 % of chrysotile or amphibole asbestos. As a result of the continuing exposure to meteorologic influences such as acid rain, sunshine, wind, frost etc., as well as to aggressive atmospheric pollutants, the surface of asbestos cement products corrodes and weathers. Thus, cement particles and asbestos fibers are released from the surface and disperse in air and rain water. Asbestos is a carcinogenic substance, and its emissions as well as its atmospheric concentrations need to be measured.
In our first investigation (1) rather high asbestos concentrations were measured in the ambient air in the vicinity of buildings containing asbestos cement products. In a later complex investigation project (2), asbestos fiber emissions and asbestos fiber concentrations in ambient air were measured in Germany 1984-1986.

MEASUREMENTS AND ANALYSES

A special equipment and procedure have been developed for sampling of fibers released into the atmosphere: fibrous emissions are sampled on Nuclepore or Membrane filters and the number of fibers, their size distributions and identities are evaluated by means of electron microscopy (SEM and TEM). Equipment and procedure have already been described elsewhere (3).
The single fibers were identified by energy dispersive x-ray analysis (EDXA). Besides, bulk analysis was provided by means of the x-ray fluorescence spectroscopy (XFS).
Crystallographic changes in weathered asbestos fibers were evaluated by the application of electron microdiffraction analysis (SAED) of single fibers (4).

RESULTS AND DISCUSSIONS

The investigations of the asbestos cement products (asbestos sheets: roofs and facades) realized during 1984 and 1986 in Germany have shown:

1. The asbestos cement surfaces corrode and weather due to the influence of atmospheric aggressive pollution (gases - SO_2, aerosols and acid rain).

2. The surface cement matrix of the material is destroyed, and a thin (app. 0.1 to 0.3 mm) layer of free deposited asbestos fibers is built up (Fig. 1).

3. The corrosion rate strongly depends on the rain acidity and on the concentration of air pollutants.

Figure 1: Photograph of a corroded and weathered asbestos cement shingle roof and electron micrograph (below) of a corroded AC-surface.

4. The wind disperses fibers into ambient air. Fiber emission factor in the range of 10^6 to 10^9 asbestos fibers per m^2/h have been measured (Fig. 2). As mentioned, the influence of corrosion time, pollution intensity and weather conditions on the fiber release have also been observed.

5. Preliminary evaluations have shown that about 20 % of free asbestos fibers are dispersed into the ambient air and 80 % are washed out by rain water.

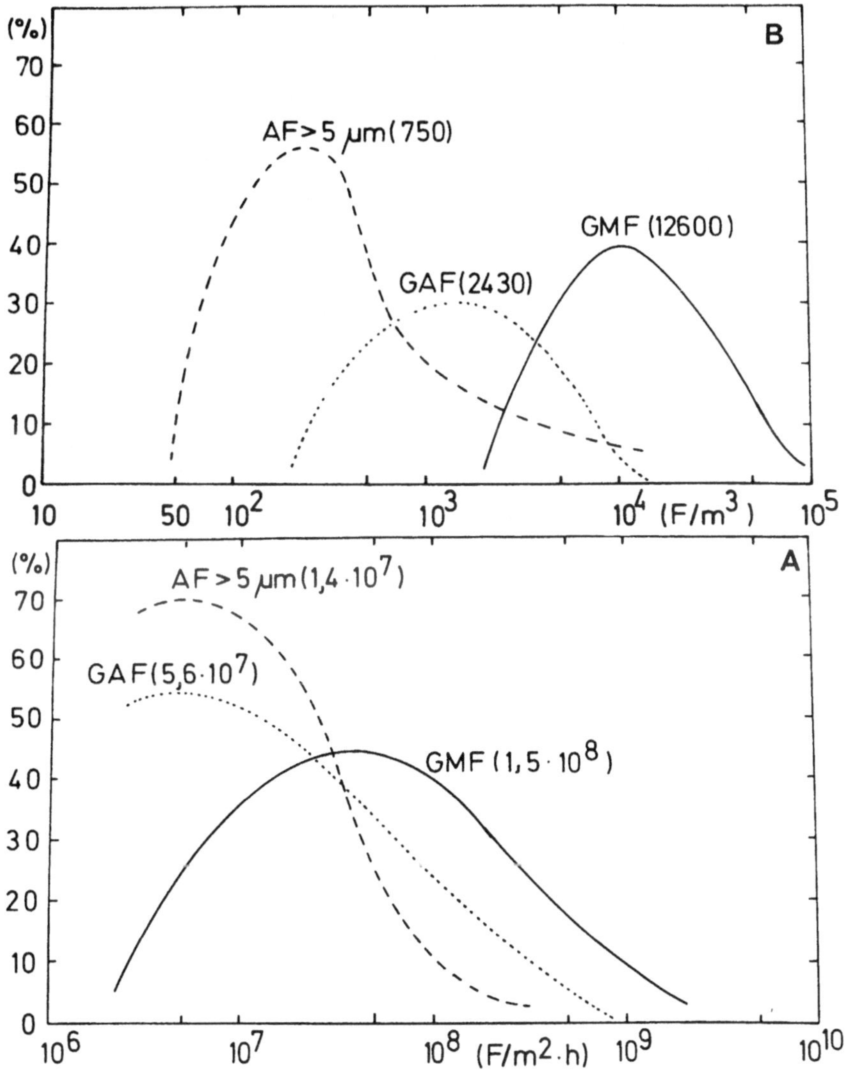

Figure 2: Schematic picture of the distribution of measured emission factors (A) and asbestos fiber concentrations (B). AF means asbestos fibers longer than 5 µm, GAF all asbestos fibers and GMF all measured mineral fibers.

6. By analysis of bulk samples as well as by analysis of individual fibers, chemical and crystallographical changes in the corroded chrysotile fibers have also been identified. Particulate air pollutants (metal compounds and organic substances) are deposited on free asbestos fibers in the corroded layer of asbestos cement products.

7. Intraperitoneal application of the released fibers from corroded asbestos cement products to animals have shown, that their carcinogenic potency is approximately as high as the potency of standardized UICC-chrysotile fibers (2).

8. Measurements of asbestos fiber concentrations in the vicinity of buildings containing corroded and weathered asbestos cement products have shown asbestos fiber concentrations (for fibers longer than 5 µm) in the range of 200 to 1200 fibers per m^3 air (Fig. 2 B). A proposed preventive limit for asbestos fiber concentrations in ambient air is considered to be 400 fibers/m^3 (5).

ACKNOWLEDGEMENTS

We are grateful to the Umweltbundesamt in Berlin for the interest in these investigations and for financial support.

REFERENCES

1. Spurny, K., G. Weiss and H. Opiela (1979) On the asbestos emissions from weathered AC-products, Staub-Reinhalt.Luft. 39, 422-427.
2. Spurny, K.R., H. Marfels, F. Pott and H. Muhle (1986) Investigations on corrosion and weathering of asbestos cement products as well as on the carcinogenic effect of the weathering products. Report UBA 104 08 314, 1-172.
3. Spurny, K.R., F.-J. Mönig and D. Hochrainer (1985) On the measurement of toxic emissions from surface sources, Staub-Reinhalt.-Luft 45, 328- 330.*)
4. Spurny, K.R. (Ed.) (1986) Physical and chemical characterization of individual airborne particles. Ellis Horwood series in analytical chemistry. J. Wiley and Sons, Chichester, 1-148.
5. Dobbertin, S. (1987) New aspects to define air quality guidelines for carcinogenic substances in the Federal Republik of Germany. This monograph.

*) The equipment (3) is commercially available at the Comp. Dr. G. Wazau, Kepler Str. 2, D-1000 Berlin 10, FRG.

ENVIRONMENTAL AND HEALTH CONTROL - ROLE OF METAL RESISTANCE IN BACTERIA

Thriene, B., A. Hellwig, K.-H. Weege, and S. Schulz
Bezirks-Hygieneinspektion und -institut Magdeburg, Wallonerberg 2-3, 3040 Magdeburg, GDR

INTRODUCTION

Bacteria have evolved a variety of means, generally referred to as resistance mechanisms, for dealing with different toxic heavy metals. These resistance mechanisms are highly specific and are determined frequently by extrachromosomal plasmids or transposons (1,2).
Plasmid borne resistance to heavy metals has been described in a great number of grampositive and gramnegative bacteria of clinical or environmental origin.
The mechanism of resistance is, on the one hand, essential for the survival of the microorganisms in metal polluted ecosystems. On the other hand, the microbial detoxification and biotransformation of heavy metals may play an important role in the circulation of toxic elements in the ecosphere.
Resistance to mercury compounds is the most intensively studied of the plasmid-borne resistances to toxic metals.
The mechanism of resistance to mercury and organomercury is based on the transformation of the compounds mentioned above to elementary volatile mercury by the enzymes reductase and lyase (3).
In numerous reports, relations between the environmental pollution by mercury and the occurrence of mercury resistant bacteria are stressed (4,5). However, no data have been reported in the literature on the prevalence of mercury resistant bacteria in the healthy population and in individuals occupationally exposed to mercury. Therefore, we looked for relations between exposure of man to mercury and the frequency of mercury resistant bacteria colonizing the human body, and for plasmids possibly involved.

MATERIALS AND METHODS

The selection of groups of individuals was based on the data on environmental contamination by heavy metals obtained in the frame of surveillance programmes. The normal population was represented by 687 individuals. In all of them occupational exposure to mercury was ruled out. The groups of individuals occupationally exposed to mercury consisted of 62 workers of different plants. The mercury contamination of the air in these plants was found at times to exceed considerably the official limits.
The isolation of mercury resistant bacteria was done on modified Endo medium containing 13.6 mg $HgCl_2$/l (for Enterobacteriaceae) and on an agar medium with a low mercury binding capacity (L 4 medium) containing 8.16 mg $HgCl_2$/l (for staphylococci). Selected mercury resistant bacteria were analysed for plasmid DNA by agarose gel electrophoresis.

RESULTS

Table 1 shows the prevalence of mercury resistant bacteria in the human population. 40.5 % of the individuals representing the normal population were found to have mercury resistant Enterobacteriaceae in their intestinal flora.

Table 1 Occurrence of mercury resistant Enterobacteriaceae in the intestinal flora

groups	% of individuals with Hg-resistance	Hg-monoresistance
individuals without occupational exposure to mercury	40.5	27.4
workers exposed to organomercurials	89.1	81.1
workers exposed to metallic mercury	35.5	23.5

According to results of our investigations the cause of this rather high prevalence might be the intake of mercury resistant bacteria via foods as well as the mercury content of several food-stuffs. Since the resistance of certain species of bacteria is mediated by the same plasmids that encode for antibiotic resistance, coselection may also be the cause of the widespread resistance to mercury.
A distinctly higher frequency of 89.1 % was determined among workers of the plant producing seed dressings. Similar results were obtained when only the monoresistance to mercury was evaluated.
Though the air in the work-rooms of the chlorine producing plant contained as much mercury as the air in the plant producing seed dressings, the prevalence of mercury resistant bacteria found in the staff of the chlorine factory did not differ very much from that of the normal population. The reason why may be the kind of mercury compounds the employees were exposed to. While only metallic mercury was found in the chlorine plant the staff of the plant producing seed dressings dealt predominantly with organic mercury compounds. We therefore conclude that only these induce mercury resistance in the human microflora (6,7). If this is true the bacteria selected should possess the lyase-reductase system detoxifying organic mercury compounds by transformation into elementary mercury. This was confirmed by the finding of a plasmid belonging to the incompatibility group H 2 in Enterobacteriaceae strains, mostly in E. coli cultured from the workers of the plant producing seed dressings. This plasmid has a molecular weight of 160 MD and encodes for an organomercury lyase, a mercury chlo-

ride reductase, and for resistance to tellurite. It was only very rarely detected in isolates from other sources. Figure 1 shows an agarose gel electrophoresis of plasmid DNA of eight mercury-resistant strains from different origin (lanes 5-12) and the reference plasmids (lanes 1-4). Lane 5 contains the DNA lysate of an E. coli strain from a worker of the plant producing seed dressings which harboured the H 2 plasmid. Mercury resistance of strains isolated from other sources is referred to plasmids with incompatibility groups F I, F II, C, I, and others (see lanes 6-12).

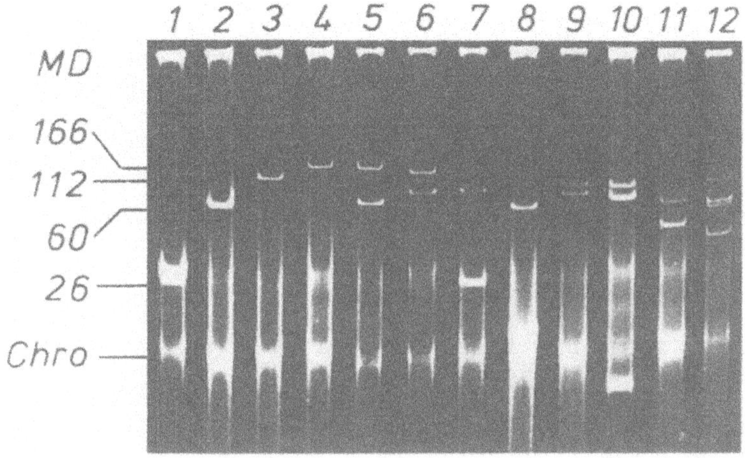

Figure 1 Agarose gel electrophoresis of plasmid DNA of mercury resistant strains of different origin

Organic mercury compounds may also be responsible for the mercury resistance of staphylococci. In the GDR mercury resistance of Staphylococcus aureus is almost exclusively met in strains of hospital origin (8). Resistant staphylococci including coagulase-negative species are only seldom isolated from healthy humans and domestic animals. In contrast to this about half of the nasal swabs taken from the employees dealing with mercury stains contained mercury resistant staphylococci which were differing phenotypically and genotypically from the hospital strains. This was discovered in hybridization tests. The strains did not carry plasmids, their mer-sequence was localized on the chromosome.

SUMMARY AND CONCLUSIONS

It was shown, that exposure to organic mercury selects mercury-resistant Enterobacteriaceae und staphylococci of

human origin. In this case bacteria with plasmid borne or chromosomally determined mercury resistance were found to be a good marker to indicate metal exposure.
Resistances are known also to other toxic metals including arsenic, cadmium, chromium, copper, silver, and zinc. Metal resistant bacteria were found both in human and in environmental material. It is suggested that patterns of bacterial resistance to toxic metals could be useful biomonitors to assess the degree of pollution and bioavailability of heavy metals in the environment. In addition, the application of specific metal resistant strains could be a practicable and economical alternative to clean up metal pollution problem.

REFERENCES

(1) Silver, S. and T.K.Misra (1984) Bacterial transformations of and resistances to heavy metals. In Genetic Control of Environmental Pollutants, eds. G.S.Omenn and A.Hollaender, Plenum Press, New York, 23-46.
(2) Summers, A.O. (1985) Bacterial resistance to toxic elements, Trends in Biotechnology, 3, 122-125.
(3) Robinson, J.B. and O.H.Tuovinen (1984) Mechanisms of microbial resistance and detoxification of mercury and organomercury compounds: Physiological, biochemical and genetic analyses, Microbiol. Rev., 48, 95-124.
(4) Khesin, R.B. and E.V.Karasyova (1984) Mercury-resistant plasmids in bacteria from a mercury and antimony deposit area, Mol.Gen.Genet., 197, 280-285.
(5) Timoney, J.F., J.Port, J.Giles, and J.Spanier (1978) Heavy-metal and antibiotic resistance in the bacterial flora of sediments of New York bight, Appl.Environ.Microbiol., 36, 465-472.
(6) Hellwig, A. und B.Thriene (1986) Zusammenhänge zwischen der Umweltbelastung mit Quecksilber und der Prävalenz und Inzidenz von Resistenzen bei Enterobacteriaceae und Staphylokokken, Med.Diss.B, Medizinische Akademie Magdeburg.
(7) Thriene, B. und A.Hellwig (1987) Die Quecksilberresistenz bei Bakterien - ein Bioindikator für eine Umweltbelastung durch Quecksilberverbindungen, VDI-Kolloquium Bioindikation, Freising 1986. In: VDI-Berichte, VDI-Verlag Düsseldorf, 609, 189-203.
(8) Witte, W., Nguyen Van Dip und R.Hummel (1980) Resistenz gegen Quecksilber und Cadmium bei Staphylococcus aureus unterschiedlicher ökologischer Herkunft, Z.Allg.Mikrobiol., 20, 517-521.

Subject Index

www.ingramcontent.com/pod-product-compliance
Lightning Source LLC
Chambersburg PA
CBHW061208240926
51846CB00006B/1150
* 9 7 8 3 6 4 2 7 3 7 6 8 8 *